The Human Brain during the Third Trimester 260- to 270-mm Crown-Rump Lengths

This twelfth of 15 short atlases reimagines the classic 5 volume *Atlas of Human Central Nervous System Development*. This volume presents serial sections from specimens between 260 mm and 270 mm with detailed annotations. An introduction summarizes human CNS developmental highlights around 6.5 months of gestation. The Glossary (available separately) gives definitions for all the terms used in this volume and all the others in the *Atlas*.

Key Features

- Classic anatomical atlases

- Detailed labeling of structures in the developing brain offers updated terminology and the identification of unique developmental features, such as germinal matrices of specific neuronal populations and migratory streams of young neurons

- Appeals to neuroanatomists, developmental biologists, and clinical practitioners

- A valuable reference work on brain development that will be relevant for decades

ATLAS OF
HUMAN CENTRAL NERVOUS SYSTEM DEVELOPMENT
Series

Volume 1: The Human Brain during the First Trimester 3.5- to 4.5-mm Crown-Rump Lengths

Volume 2: The Human Brain during the First Trimester 6.3- to 10.5-mm Crown-Rump Lengths

Volume 3: The Human Brain during the First Trimester 15- to 18-mm Crown-Rump Lengths

Volume 4: The Human Brain during the First Trimester 21- to 23-mm Crown-Rump Lengths

Volume 5: The Human Brain during the First Trimester 31- to 33-mm Crown-Rump Lengths

Volume 6: The Human Brain during the First Trimester 40- to 42-mm Crown-Rump Lengths

Volume 7: The Human Brain during the First Trimester 57- to 60-mm Crown-Rump Lengths

Volume 8: The Human Brain during the Second Trimester 96- to 150-mm Crown-Rump Lengths

Volume 9: The Human Brain during the Second Trimester 160- to 170-mm Crown-Rump Lengths

Volume 10: The Human Brain during the Second Trimester 190- to 210-mm Crown-Rump Lengths

Volume 11: The Human Brain during the Third Trimester 225- to 235-mm Crown-Rump Lengths

Volume 12: The Human Brain during the Third Trimester 260- to 270-mm Crown-Rump Lengths

Volume 13: The Human Brain during the Third Trimester 310- to 350-mm Crown-Rump Lengths

Volume 14: The Spinal Cord during the First Trimester

Volume 15: The Spinal Cord during the Second and Third Trimesters and the Early Postnatal Period

The Human Brain during the Third Trimester 260- to 270-mm Crown-Rump Lengths

Atlas of Human Central Nervous System Development, Volume 12

Shirley A. Bayer

Joseph Altman

CRC Press

Taylor & Francis Group

Boca Raton London New York

CRC Press is an imprint of the
Taylor & Francis Group, an **informa** business

Designed cover: Shirley A. Bayer and Joseph Altman

First edition published 2024
by CRC Press
2385 NW Executive Center Drive, Suite 320, Boca Raton, FL 33431

and by CRC Press
4 Park Square, Milton Park, Abingdon, Oxon, OX14 4RN

CRC Press is an imprint of Taylor & Francis Group, LLC

LCCN no. 2022008216

ISBN: 978-1-032-22878-5 (hbk)
ISBN: 978-1-032-22876-1 (pbk)
ISBN: 978-1-003-27461-2 (ebk)

DOI: 10.1201/9781003274612

Typeset in Times Roman
by KnowledgeWorks Global Ltd.

Publisher's note: This book has been prepared from camera-ready copy provided by the authors.
Access the Support Material: www.routledge.com/9781032228785

CONTENTS

ACKNOWLEDGMENTS

We thank the late Dr. William DeMyer, pediatric neurologist at Indiana University Medical Center, for access to his personal library on human CNS development. We also thank the staff of the National Museum of Health and Medicine that were at the Armed Forces Institute of Pathology, Walter Reed Hospital, Washington, D.C. when we collected data in 1995 and 1996: Dr. Adrianne Noe, Director; Archibald J. Fobbs, Curator of the Yakovlev Collection; Elizabeth C. Lockett; and William Discher. We are most grateful to the late Dr. James M. Petras at the Walter Reed Institute of Research who made his darkroom facilities available so that we could develop all the photomicrographs on location rather than in our laboratory in Indiana. Finally, we thank Chuck Crumly, Neha Bhatt, Kara Roberts, Michele Dimont, and Rebecca Condit for expert help during production of the manuscript.

AUTHORS

Shirley A. Bayer received her PhD from Purdue University in 1974 and spent most of her scientific career working with Joseph Altman. She was a professor of biology at Indiana-Purdue University in Indianapolis for several years, where she taught courses in human anatomy and developmental neurobiology while continuing to do research in brain development. Her lengthy publication record of dozens of peer-reviewed, scientific journal articles extends back to the mid 1970s. She has co-authored several books and many articles with her late spouse, Joseph Altman. It was her research (published in *Science* in 1982) that proved that new neurons are added to granule cells in the dentate gyrus during adult life, a unique neuronal population that grows. That paper stimulated interest in the dormant field of adult neurogenesis.

Joseph Altman, now deceased, was born in Hungary and migrated with his family via Germany and Australia to the US. In New York, he became a graduate student in psychology in the laboratory of Hans-Lukas Teuber, earning a PhD in 1959 from New York University. He was a postdoctoral fellow at Columbia University, and later joined the faculty at the Massachusetts Institute of Technology. In 1968, he accepted a position as a professor of biology at Purdue University. During his career, he collaborated closely with Shirley A. Bayer. From the early 1960s-2016, he published many articles in peer-reviewed journals, books, monographs, and free online books that emphasized developmental processes in brain anatomy and function. His most important discovery was adult neurogenesis, the creation of new neurons in the adult brain. This discovery was made in the early 1960s while he was based at MIT, but was largely ignored in favor of the prevailing dogma that neurogenesis is limited to prenatal development. After Dr. Bayer's paper proved new neurons are added to granule cells in the hippocampus, Dr. Altman's monumental discovery became more accepted. During the 1990s, new researchers "rediscovered" and confirmed his original finding. Adult neurogenesis has recently been proven to occur in the dentate gyrus, olfactory bulb, and striatum through the measurement of Carbon-14—the levels of which changed during nuclear bomb testing throughout the 20th century—in postmortem human brains. Today, many laboratories around the world are continuing to study the importance of adult neurogenesis in brain function. In 2011, Dr. Altman was awarded the Prince of Asturias Award, an annual prize given in Spain by the Prince of Asturias Foundation to individuals, entities, or organizations globally who make notable achievements in the sciences, humanities, and public affairs. In 2012, he received the International Prize for Biology - an annual award from the Japan Society for the Promotion of Science (JSPS) for "outstanding contribution to the advancement of research in fundamental biology." This Prize is one of the most prestigious honors a scientist can receive. When Dr. Altman died in 2016, Dr. Bayer continued the work they started over 50 years ago. In her late husband's honor, she created the Altman Prize, awarded each year by JSPS to an outstanding young researcher in developmental neuroscience.

INTRODUCTION

A. Specimens and Organization

Volume 12 in the Atlas Series presents the human brain in three normal specimens from the Yakovlev Collection[1] at 7.5 to 8 months during the third trimester. These specimens were analyzed in Volume 2 of the original *Atlas of Human Central Nervous System Development* (Bayer and Altman, 2004). Most fetuses are viable *ex utero* at this time. Nearly all the structures present in the adult brain are recognizable and are maturing from the diencephalon to the medulla. But remnants of the embryonic nervous system remain in the cerebral cortex and the cerebellar cortex.

This volume contains serial grayscale photographs of Nissl-stained sections of a sagittal specimen (Y15-60, **Part II**), a frontal specimen (Y14-59, **Part III**), and a horizontal specimen (Y187-65, **Part IV**) with crown-rump (CR) lengths of 260 to 270 mm. All specimens are in the 30th to 32nd gestational weeks (GW). **Sagittal plates** are ordered from medial to lateral; the anterior part of each photographed section is facing left, posterior right. **Frontal plates** are presented from anterior (first) to posterior (last); the dorsal part of each section is at the top, the ventral part at the bottom; the midline is in the vertical center. **Horizontal plates** are presented from dorsal (first) to ventral

1. The *Yakovlev Collection* (designated by a **Y** prefix in the specimen number) is the work of Dr. Paul Ivan Yakovlev (1894–1983), a neurologist affiliated with Harvard University. Throughout his career, Yakovlev collected many diseased and normal human brains. He invented a giant microtome that was capable of sectioning entire human brains. Later, he became interested in the developing brain and collected many during the second and third trimesters. The normal brains in the developmental group were cataloged by Haleem (1990) and were examined by us during 1996 and 1997. The collection was moved to the National Museum of Health and Medicine when the Armed Forces Institute of Pathology (AFIP) closed at Walter Reed Hospital and is still available for research.

(last); the anterior part of each section faces left, the posterior part faces right; the midline is in the horizontal center. Each **plate** is in two parts: **A**, on the left, shows the full-contrast photograph without labels; **B,** on the right, shows a low-contrast copy of each photograph with unabbreviated labels. For each specimen, a series of serially spaced **low-magnification plates** show the entire section to identify large structures. The brain core is shown in many **high-magnification plates** to identify smaller structures. In addition, several **very-high-magnification plates** show the cerebral and cerebellar cortices. Because our emphasis is on development, transient structures that appear only in immature brains are labeled in *italics*, either directly in some of the high-magnification plates or in **bold numbers** that refer to labels in a list. During dissection, embedding, cutting, and staining, some of the sections illustrated were torn; that damage is sometimes surrounded by *dashed lines*.

B. Developmental Highlights

Figures 1-2 compare late second trimester, early third trimester, and middle third trimester brains in horizontal sections of the dorsal telencephalon. The overall growth of the brain is obvious when the sections are stacked in a "layer cake" arrangement (**Fig. 1**). The side by side comparison (**Fig. 2**) shows that the cortical plate gradually increases its thickness and rapidly grows lengthwise. The white matter (*pale yellow*) greatly increases its thickness in most cortical areas except the insula. On the other hand, layers 3-6 of the *stratified transitional field* slow their growth. Cortical gyrification continues to elaborate as more fibers enter, and the white matter increases its relative volume. Fibers come from the thalamus (via the enlarging internal capsule) and from the opposite cortex (via the expanding corpus callosum). In addition, associational fibers are accumulating in ipsilat-

eral sides of the cortex to establish connections with neurons in different cortical areas.

Studies using magnetic resonance imaging shed more light on the development of third trimester brains. Garcia et al. (2018) performed a longitudinal study imaging the brains of the same preterm infants at two- to three-week intervals that were born from 28 to 38 gestational weeks. They found that early sulci and gyri are highly conserved and show little individual variability. On the other hand, secondary and tertiary gyri that develop during the third trimester and up to the young adult period are more variable. They also analyzed the rate of growth of different cortical areas and found that primary sensory cortical areas (somatosensory, visual) and the insula expand early then slow their expansion rate compared to the rest of the cortex. We suggest that this growth pattern is due to the "anchoring" by specific contacts with thalamic axons (Altman and Bayer, 2015; Van Essen, 1997). In contrast, association areas in the frontal, parietal, and temporal lobes increase their rate of expansion during the third trimester and into the postnatal period (*see* Figs. 5-7 in Garcia et al., 2018).

Wilson et al., (2021) report that organized fiber tracts are detectable in the white matter of fetal brains during the second and third trimesters (22 to 37 gestational weeks). They could confirm the trajectory of the corticospinal tract from the white matter in the paracentral lobule all the way into the brainstem. Furthermore, the genu and splenium of the corpus callosum showed different maturational patterns. They could also detect the visual radiation from the thalamus to the occipital lobe. Thus, the increased white matter that we see in the specimens in this volume is not just an unorganized space, but instead is a highly dynamic region where axons are fasciculating and maintaining topographical order. It is well known

**Superimposed Third Trimester
and Second Trimester Brains
in the Horizontal Plane**

**Bottom layer: GW 30
Middle layer: GW 26
Top layer: GW 23**

*NEP- neuroepithelium
STF-stratified transitional field
SVZ-subventricular zone*

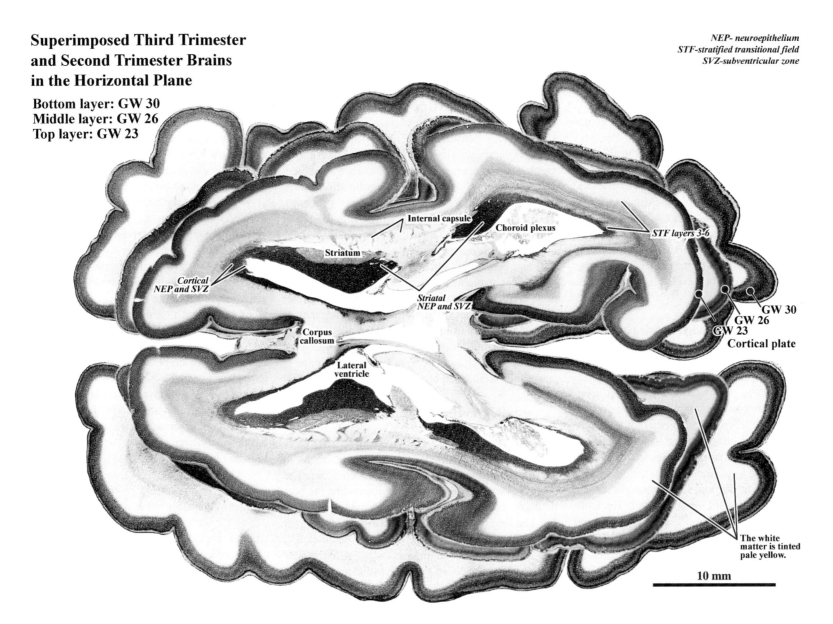

Internal capsule

Choroid plexus

STF layers 3-6

Striatum

*Cortical
NEP and SVZ*

*Striatal
NEP and SVZ*

GW 30

GW 26

GW 23

Cortical plate

**Corpus
callosum**

**Lateral
ventricle**

**The white
matter is tinted
pale yellow.**

10 mm

that the fibers going to the cortex from relay nuclei in the thalamus are topographically organized. Thus, as the connections are made, order prevails from the very start of cortical circuitry.

Doria et al., (2010) analyzed system-state brain networks with functional MRI techniques. During the middle third trimester, several circuits involving the cerebral cortex are presumed to be functional: the medial visual circuit around the calcarine sulcus, the auditory circuit in the temporal lobe, and the somato-motor circuit around the central sulcus. In addition, there are functional circuits in the thalamus, cerebellum, and throughout the brainstem.

Side-by-Side Comparison of Third Trimester and Second Trimester Brains in the Horizontal Plane

Growth related to increased gyrification:

Cortical plate increases thickness and linear surface

White matter increases depth in most areas (insula excepted)

STF layers 4-6 decrease in relative size

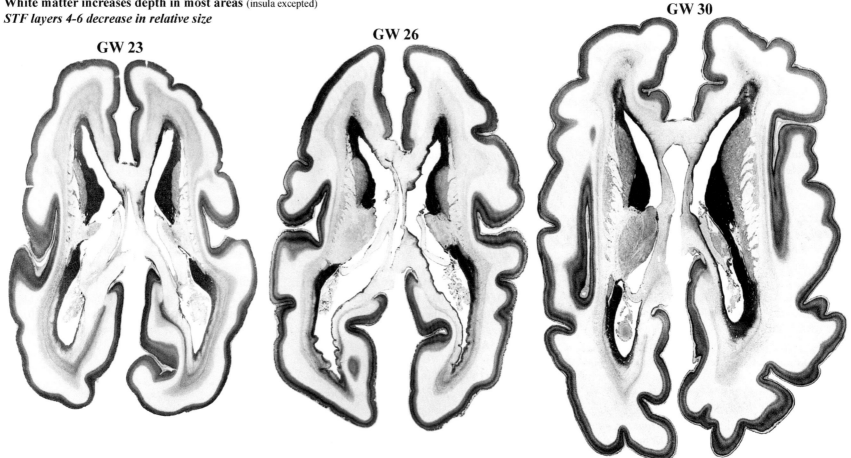

GW 30

GW 26

GW 23

10 mm

Figures 1 and 2 (*facing pages*). A comparison of one second trimester brain (GW 23, *top and left*), an early third trimester brain (GW 23, *middle*), and a middle third trimester brain (GW 30, *bottom and right*) cut the horizontal plane. All brains are shown at the same scale to emphasize the relative growth of different components of the cerebral cortex. Note that the internal capsule and corpus callosum continue to thicken because many new axons enter. Massive numbers of axons invade the white matter (*pale yellow*). These axons pile up in the white matter and their terminal branches grow into the cortical plate, causing it to thicken and lengthen as synaptic contacts are beginning to be established. (GW 23 is shown in Plate 1, Volume 10, Bayer and Altman, in Press; GW 26 in Plate 28, Volume 11, Bayer and Altman, in Press; GW 30 in Plate 58, this volume).

REFERENCES

Altman J, Bayer SA. (2015) *Development of the Human Neocortex*. Ocala, FL, Laboratory of Developmental Neurobiology, neurondevelopment.org.

Bayer SA, Altman J (2004) *Atlas of Human Central Nervous System Development*, Volume 2: *The Human Bran during the Third Trimester.* Boca Raton, FL, CRC Press.

Bayer SA, Altman J (in press) *The Human Brain during the Second Trimester 190- to 210-mm Crown-Rump Lengths, Atlas of Human Central Nervous System Development*, Volume 10. Taylor and Francis, CRC Press.

Bayer SA, Altman J (in press) *The Human Brain during the Second Trimester 225- to 235-mm Crown-Rump Lengths, Atlas of Human Central Nervous System Development*, Volume 11. Taylor and Francis, CRC Press.

Curtis BA, Jacobson S, Marcus EM (1972) *An Introduction to the Neurosciences*, Philadelphia: W. B. Saunders.

Doria V, Beckmann CF, Arichi T, et al. (2010) Emergence of resting state networks in the preterm human brain. *Proceedings of the National Academy of Sciences*, 107 (46) 20015-20020.

Garcia KE, Robinson EC, Alexopoulos D, et al. (2018) Dynamic patterns of cortical expansion during folding of the preterm human brain. *Proceedings of the National Academy of Sciences*, 115:3156-3161.

Haleem M (1990) *Diagnostic Categories of the Yakovlev Collection of Normal and Pathological Anatomy and Development of the Brain.* Washington, D.C. Armed Forces Institute of Pathology.

Larroche JC (1966) The development of the central nervous system during intrauterine life. In: *Human Development*, F. Falkner (ed.), Philadelphia: W. B. Saunders, pages 257-276.

Van Essen DC (1997) A tension-based theory of morphogenesis and compact wiring in the central nervous system. *Nature*, 385:313-318.

Wilson S, Pietsch M, Cordero-Grande L, et al. (2021) Development of human white matter pathways in utero over the second and third trimester. *Proceedings of the National Academy of Sciences*, 118 (20) e2023598118.

PART II: Y15-60
CR 270 mm (GW 32)
Sagittal

This specimen is case number BX-15-60 (Perinatal RPSL) in the Yakovlev Collection. A female infant survived for one hour after a premature birth. Death occurred because a hyaline membrane obstructed the airway to the lungs. The brain was cut in the sagittal plane in 35-μm thick sections and is classified as a Normative Control in the Yakovlev Collection (Haleem, 1990). Since there is no photograph of this brain before it was embedded and cut, the photograph of the medial view of another GW 32 brain that Larroche published in 1966 (**Figure 3**) is used.

Photographs of 6 Nissl-stained low-magnification sections are shown in **Plates 1-6**. Higher-magnification views of the brain core cerebellum are shown in **Plates 7-12**. Very high-magnification views of different regions of the cerebellar cortex are shown in **Plates 13-16**. Because the section numbers decrease from **Plate 1** (most medial) to **Plate 6** (most lateral), they are from the left side of the brain; the right side has higher section numbers proceeding medial to lateral. The cutting plane of this brain is nearly parallel to the midline in anterior and posterior parts of each section. However, the occipital lobe has been displaced toward the left. For example, the occipital lobe in **Plate 1** is from the right side of the brain. There is no occipital lobe in **Plate 2**, and the left occipital lobe first appears in **Plate 3**. The sections chosen for illustration are spaced closer together near the midline to show small structures in the diencephalon, midbrain, pons, and medulla.

Y15-60 contains a dwindling number of immature features. In cortical regions of the telencephalon, remnants of the germinal matrices are present in all lobes of the cerebral cortex where the **neuroepithelium/subventricular zones** are presumably generating neocortical interneurons and glia. Migrating and sojourning neurons and/or glia are visible in all lobes of the cerebral cortex in less prominent **stratified transitional fields**, thin in the occipital lobe, and thicker in the frontal, parietal, and temporal lobes.

More neurons, glia, and their mitotic precursor cells are migrating through the olfactory peduncle toward the olfactory bulb (**rostral migratory stream**) from a presumed source area in the germinal matrix at the junction between the cerebral cortex, striatum, and nucleus accumbens. Within the lateral parts of the cerebral cortex, the **lateral migratory stream** contains neurons and glia that percolate through the claustrum, endopiriform nucleus, external capsule, and uncinate fasciculus. These cells appear to be heading toward the insular cortex, primary olfactory cortex, temporal cortex, and basolateral parts of the amygdaloid complex.

In the basal ganglia, there is a thick **neuroepithelium/subventricular zone** overlying the striatum and nucleus accumbens where neurons and glia are being generated; some of these, especially from the accumbal area, will enter the **rostral migratory stream**. Another region of active neurogenesis in the telencephalon is the **subgranular zone** in the hilus of the dentate gyrus that is the source of granule cells. Remnants of this germinal zone is retained throughout life as it continues to generate dentate granule cells.

The septum, fornix, and Ammon's horn have a dense layer at the ventricle. These layers are presumed to be generating glia and the ependymal lining of the ventricle, a **glioepithelium/ependyma**.

Most of the structures in the diencephalon appear to be settled and are maturing, and the third ventricle is lined by a thin **glioepithelium/ependyma**. In the midbrain and anterior pons, there is a slightly thicker and more convoluted **glioepithelium/ependyma** lining the posterior cerebral aqueduct and anterior fourth ventricle. A thin **glioepithelium/ependyma** lines the fourth ventricle in the posterior pons and anterior medulla, but that thickens in the posterior medulla.

The **external germinal layer (egl)** covers the entire surface of the cerebellar cortex and is actively producing basket, stellate, and granule cells. The **germinal trigone** is visible at the base of the nodulus and along the floccular peduncle; choroid plexus cells and glia are originating here, and it was the source of the **egl** early on.

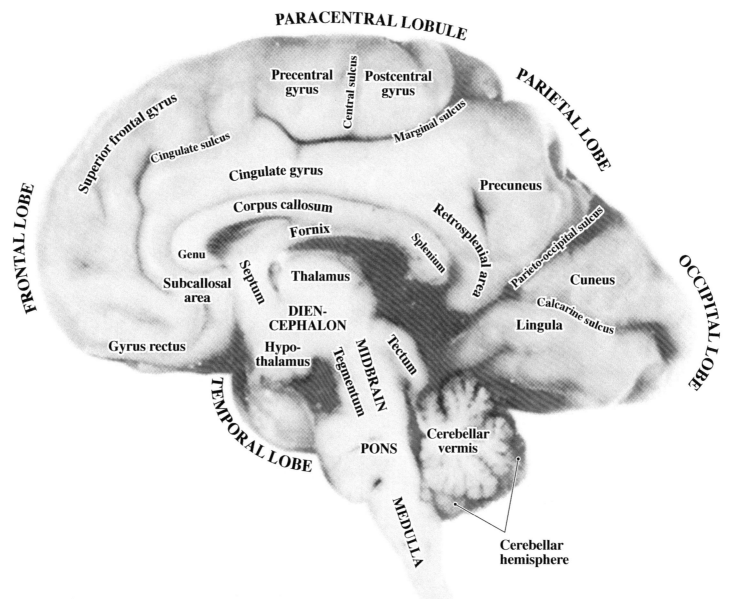

PARACENTRAL LOBULE

PARIETAL LOBE

FRONTAL LOBE

OCCIPITAL LOBE

TEMPORAL LOBE

Precentral gyrus

Postcentral gyrus

Central sulcus

Marginal sulcus

Superior frontal gyrus

Cingulate sulcus

Cingulate gyrus

Precuneus

Corpus callosum

Fornix

Retrosplenial area

Parieto-occipital sulcus

Splenium

Genu

Septum

Thalamus

Cuneus

Subcallosal area

Calcarine sulcus

DIEN-CEPHALON

Lingula

Gyrus rectus

Hypo-thalamus

Tegmentum

MIDBRAIN

Tectum

PONS

Cerebellar vermis

MEDULLA

Cerebellar hemisphere

Spinal cord

Figure 3. Midline sagittal view of a GW 32 brain with major structures in the cerebral hemispheres and brainstem labeled. (This is part of Figure 2-9 on page 27 in B. A. Curtis, S. Jacobson, and E. M. Marcus (1972) *An Introduction to the Neurosciences*, Philadelphia: W. B. Saunders. The photograph was originally published by J. C. Larroche (1966) The development of the central nervous system during intrauterine life. In: *Human Development*, F. Falkner (ed.), Philadelphia: W. B. Saunders, page 259.)

8

PLATE 1A
CR 270 mm
GW 32, Y15-60
Sagittal
Section 801

See detail of the brain core
and cerebellum in
Plates 7A and B.

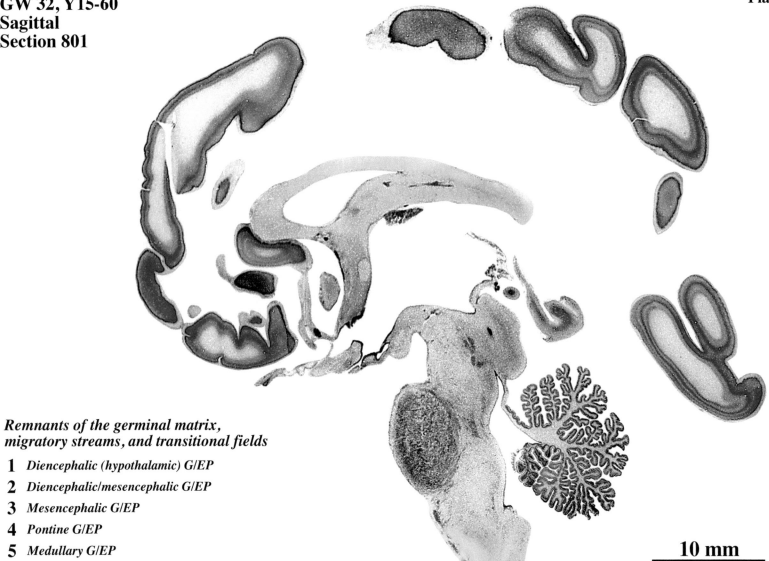

*Remnants of the germinal matrix,
migratory streams, and transitional fields*

1 *Diencephalic (hypothalamic) G/EP*

2 *Diencephalic/mesencephalic G/EP*

3 *Mesencephalic G/EP*

4 *Pontine G/EP*

5 *Medullary G/EP*

6 *Germinal trigone (cerebellum)*

7 *External germinal layer (cerebellum)*

8 *Subpial granular layer (cortical)*

G/EP - Glioepithelium/ependyma

10 mm

**High-magnification views
of the cerebellar cortex
are in Plates 13-16.**

PLATE 1B

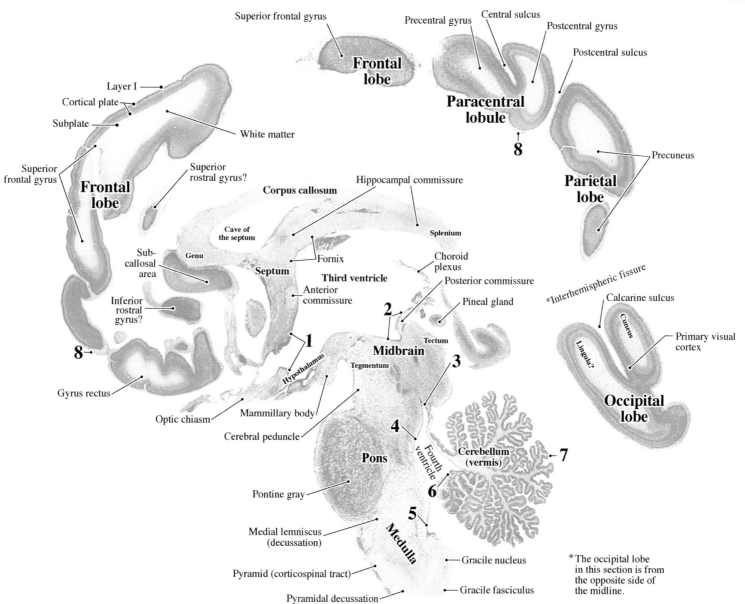

Superior frontal gyrus

Frontal lobe

Central sulcus

Precentral gyrus

Postcentral gyrus

Postcentral sulcus

Paracentral lobule

8

Layer I

Cortical plate

Subplate

White matter

Superior frontal gyrus

Frontal lobe

Superior rostral gyrus?

Corpus callosum

Hippocampal commissure

Parietal lobe

Precuneus

Sub-callosal area

Cave of the septum

Septum

Genu

Fornix

Splenium

Choroid plexus

Third ventricle

Anterior commissure

Posterior commissure

Pineal gland

*Interhemispheric fissure

Calcarine sulcus

Cuneus

Inferior rostral gyrus?

2

Midbrain

Tectum

Primary visual cortex

Lingula?

8

1

Hypothalamus

Tegmentum

3

Occipital lobe

Gyrus rectus

Optic chiasm

Mammillary body

Cerebral peduncle

4

Fourth ventricle

Cerebellum (vermis)

7

Pons

6

Pontine gray

5

Medulla

Medial lemniscus (decussation)

Gracile nucleus

Pyramid (corticospinal tract)

Gracile fasciculus

*The occipital lobe in this section is from the opposite side of the midline.

Pyramidal decussation

PLATE 2A
CR 270 mm
GW 32, Y15-60
Sagittal
Section 741

See detail of the brain core and cerebellum in Plates 8A and B.

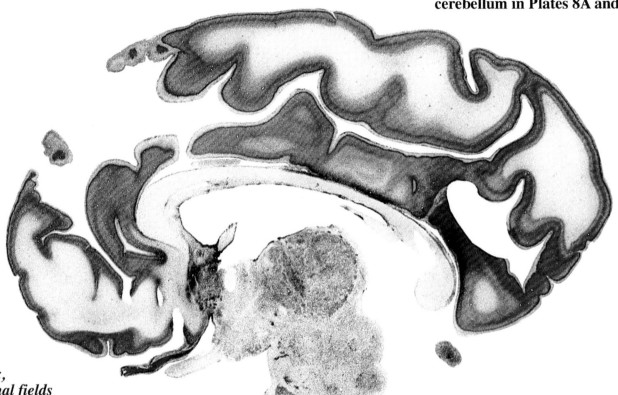

Remnants of the germinal matrix, migratory streams, and transitional fields

1 *Rostral migratory stream*

2 *Accumbent NEP and SVZ (intermingled with the source of the rostral migratory stream)*

3 *Callosal sling*

4 *Callosal GEP*

5 *Fornical GEP*

6 *Septal G/EP*

7 *Strionuclear GEP*

8 *Pontine G/EP*

9 *Germinal trigone (cerebellum)*

10 *External germinal layer (cerebellum)*

11 *Subpial granular layer (cortical)*

GEP - Glioepithelium
G/EP - Glioepithelium/ependyma
NEP - Neuroepithelium
SVZ - Subventricular zone

10 mm

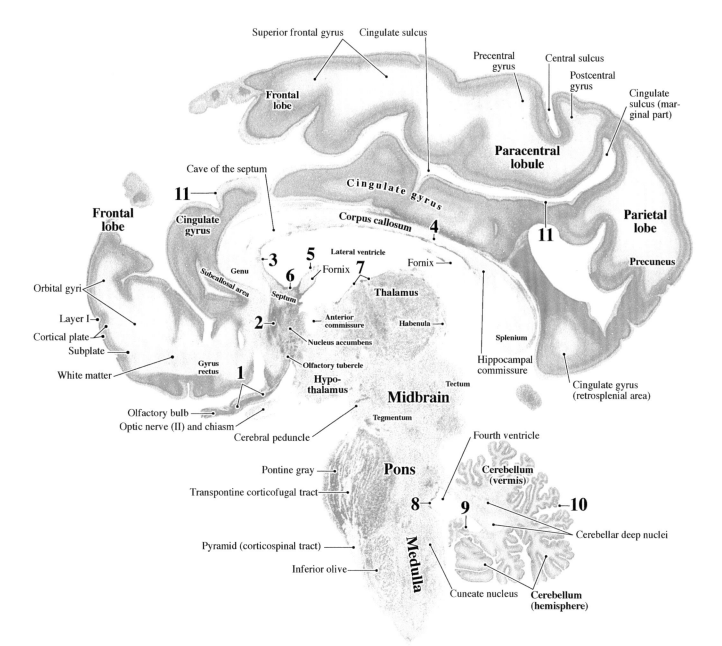

Superior frontal gyrus

Cingulate sulcus

Precentral gyrus

Central sulcus

Postcentral gyrus

Cingulate sulcus (marginal part)

Frontal lobe

Paracentral lobule

Parietal lobe

Cave of the septum

11

Cingulate gyrus

Frontal lobe

Cingulate gyrus

Corpus callosum

4

11

Lateral ventricle

Precuneus

Genu

3

5

Fornix

Subcallosal area

6

Fornix

7

Orbital gyri

Septum

Thalamus

Layer I

Anterior commissure

Habenula

Splenium

Cortical plate

2

Nucleus accumbens

Subplate

Olfactory tubercle

Hippocampal commissure

White matter

Gyrus rectus

1

Hypo-thalamus

Tectum

Cingulate gyrus (retrosplenial area)

Olfactory bulb

Midbrain

Optic nerve (II) and chiasm

Tegmentum

Cerebral peduncle

Fourth ventricle

Pontine gray

Pons

Cerebellum (vermis)

Transpontine corticofugal tract

8

9

10

Cerebellar deep nuclei

Pyramid (corticospinal tract)

Medulla

Inferior olive

Cuneate nucleus

Cerebellum (hemisphere)

12

PLATE 3A
CR 270 mm
GW 32, Y15-60
Sagittal
Section 681

See detail of the brain core and
cerebellum in Plates 9A and B.

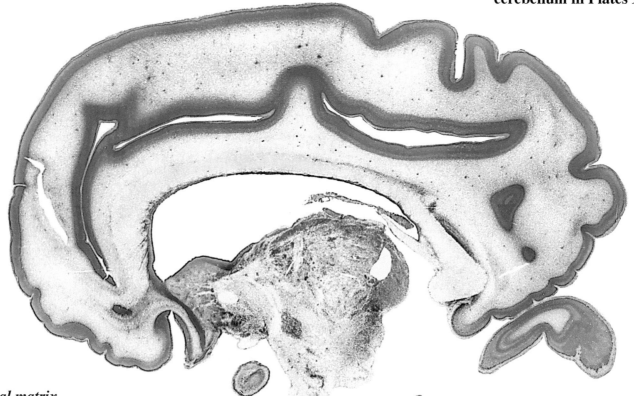

Remnants of the germinal matrix,
migratory streams, and transitional fields

1 *Rostral migratory stream*

2 *Accumbent NEP and SVZ*
   (intermingled with the source of the rostral migratory stream)

3 *Rostral migratory stream (source area)*

4 *Frontal NEP, SVZ, and STF*

5 *Callosal GEP*

6 *Fornical GEP*

7 *Anteromedial striatal NEP and SVZ*

8 *Strionuclear GEP*

9 *External germinal layer (cerebellum)*

10 *Subpial granular layer (cortical)*

GEP - Glioepithelium
NEP - Neuroepithelium
STF - Stratified transitional field
SVZ - Subventricular zone

10 mm

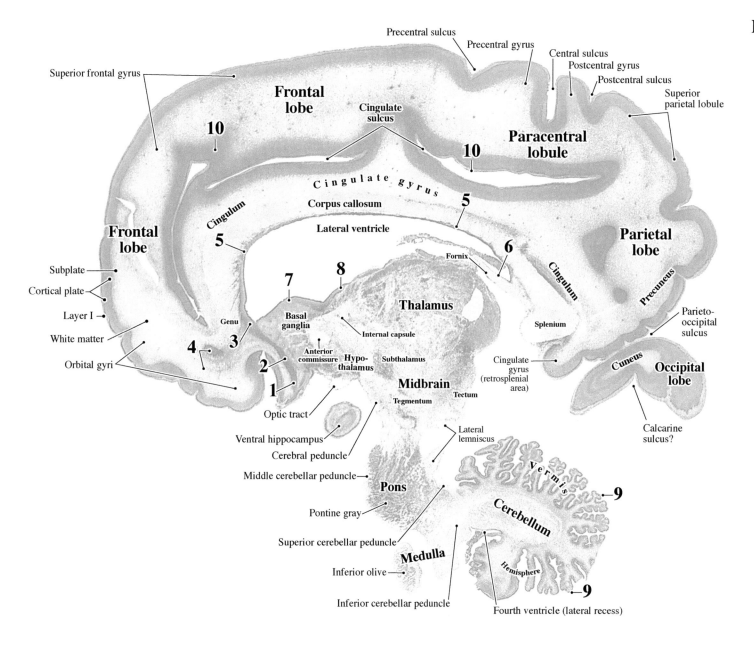

Precentral sulcus

Precentral gyrus

Central sulcus

Postcentral gyrus

Postcentral sulcus

Superior
parietal lobule

Superior frontal gyrus

**Frontal
lobe**

Cingulate
sulcus

10

10

**Paracentral
lobule**

C i n g u l a t e   g y r u s

Cingulum

5

Corpus callosum

Lateral ventricle

5

**Parietal
lobe**

**Frontal
lobe**

5

6

Fornix

Cingulum

Subplate

8

7

Thalamus

Splenium

Precuneus

Cortical plate

Genu

Basal
ganglia

Internal capsule

Cingulate
gyrus
(retrosplenial
area)

Parieto-
occipital
sulcus

Layer I

3

White matter

4

2

Anterior
commissure

Hypo-
thalamus

Subthalamus

Midbrain

Cuneus

**Occipital
lobe**

Orbital gyri

1

Tegmentum

Tectum

Optic tract

Calcarine
sulcus?

Ventral hippocampus

Cerebral peduncle

Lateral
lemniscus

V e r m i s

9

Middle cerebellar peduncle

Pons

Cerebellum

Pontine gray

Superior cerebellar peduncle

Medulla

Hemisphere

Inferior olive

9

Inferior cerebellar peduncle

Fourth ventricle (lateral recess)

PLATE 4A
CR 270 mm
GW 32, Y15-60
Sagittal
Section 581

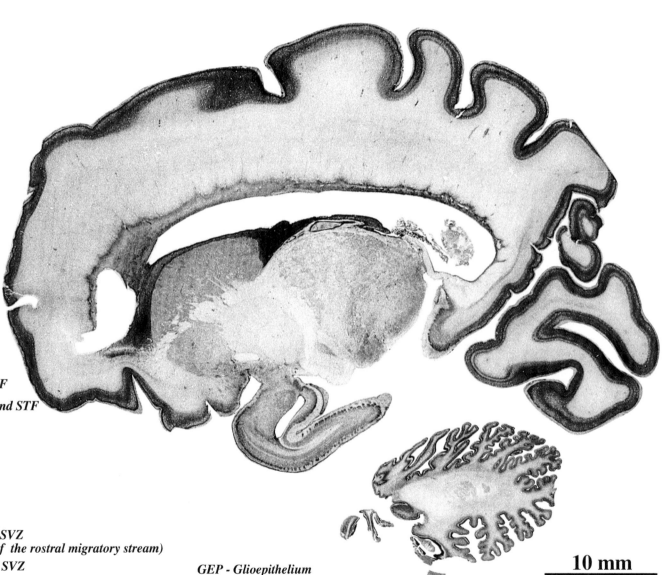

See detail of the brain core and cerebellum in Plates 10A and B.

Remnants of the
germinal matrix,
migratory streams,
and transitional fields

1   *Frontal NEP, SVZ, and STF*

2   *Paracentral NEP, SVZ, and STF*

3   *Parahippocampal NEP, SVZ, and STF*

4   *Callosal GEP*

5   *Fornical GEP*

6   *Alvear GEP*

7   *Subgranular zone (dentate)*

8   *Amygdaloid G/EP*

9   *Anterolateral striatal NEP and SVZ*
     (intermingled with the source of the rostral migratory stream)

10 *Anteromedial striatal NEP and SVZ*

11 *Strionuclear GEP*

12 *External germinal layer (cerebellum)*

13 *Subpial granular layer (cortical)*

GEP - Glioepithelium
G/EP - Glioepithelium/ependyma
NEP - Neuroepithelium
STF - Stratified transitional field
SVZ - Subventricular zone

10 mm

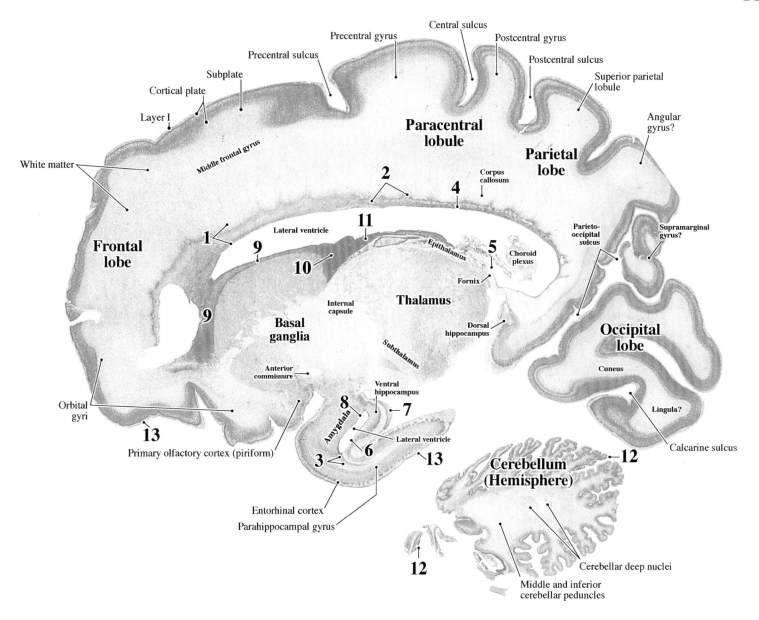

Central sulcus

Precentral gyrus

Postcentral gyrus

Precentral sulcus

Postcentral sulcus

Superior parietal lobule

Subplate

Cortical plate

Angular gyrus?

Layer I

Paracentral lobule

Parietal lobe

White matter

Middle frontal gyrus

Corpus callosum

2

4

11

Lateral ventricle

Parieto-occipital sulcus

Supramarginal gyrus?

Frontal lobe

1

9

Epithalamus

5

Choroid plexus

9

10

Fornix

Internal capsule

Thalamus

Occipital lobe

Dorsal hippocampus

9

Basal ganglia

Subthalamus

Cuneus

Anterior commissure

Ventral hippocampus

Lingula?

Orbital gyri

8

7

13

Amygdala

6

Lateral ventricle

13

Calcarine sulcus

3

12

Cerebellum (Hemisphere)

Primary olfactory cortex (piriform)

Entorhinal cortex

Parahippocampal gyrus

Cerebellar deep nuclei

12

Middle and inferior cerebellar peduncles

PLATE 5A
CR 270 mm
GW 32, Y15-60
Sagittal
Section 481

See detail of the brain core and
cerebellum in Plates 11A and B.

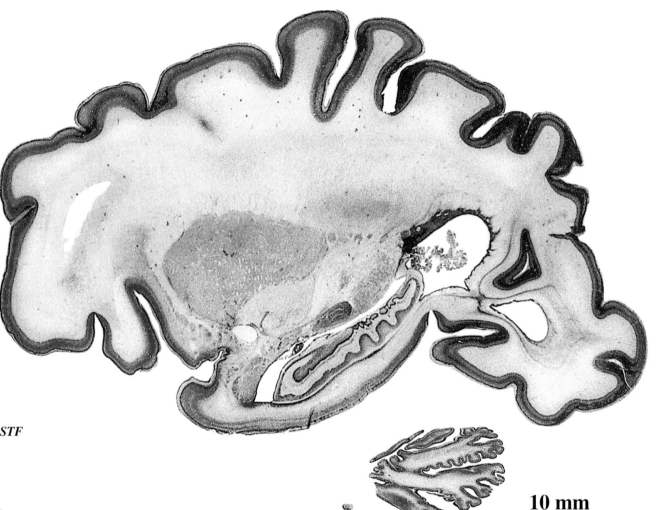

10 mm

Remnants of the
germinal matrix,
migratory streams,
and transitional fields

1 *Frontal STF*

2 *Parietal NEP, SVZ, and STF*

3 *Occipital NEP, SVZ, and STF*

4 *Parahippocampal NEP, SVZ, and STF*

5 *Fornical GEP*

6 *Alvear GEP*

7 *Subgranular zone (dentate)*

8 *Lateral migratory stream (cortical)*

9 *Amygdaloid G/EP*

10 *Posterior striatal neuroepithelium and subventricular zone*

11 *Strionuclear GEP*

12 *External germinal layer (cerebellum)*

13 *Subpial granular layer (cortical)*

GEP - Glioepithelium
G/EP - Glioepithelium/ependyma
NEP - Neuroepithelium
STF - Stratified transitional field
SVZ - Subventricular zone

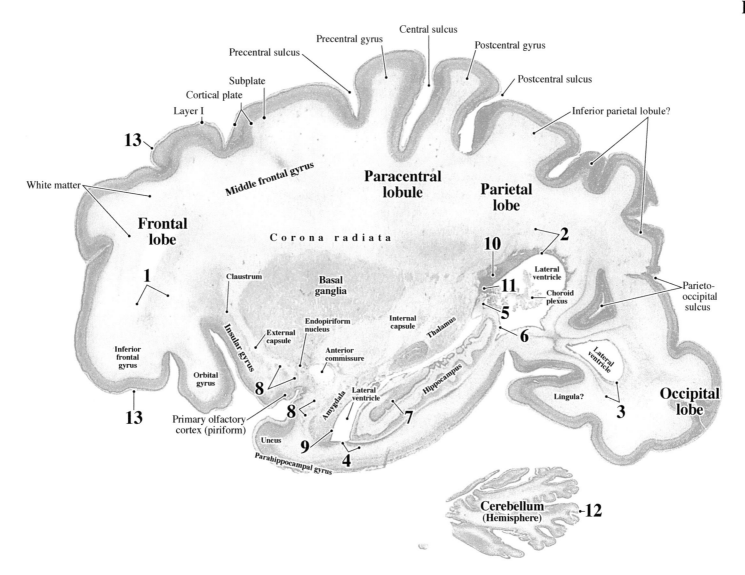

Precentral gyrus

Central sulcus

Postcentral gyrus

Precentral sulcus

Postcentral sulcus

Subplate

Inferior parietal lobule?

Cortical plate

Layer I

13

White matter

Middle frontal gyrus

Paracentral lobule

Parietal lobe

Frontal lobe

C o r o n a   r a d i a t a

2

10

1

Claustrum

Basal ganglia

Lateral ventricle

11

Choroid plexus

Parieto-occipital sulcus

Endopiriform nucleus

Internal capsule

5

Inferior frontal gyrus

External capsule

Anterior commissure

Thalamus

6

Insular gyrus

Orbital gyrus

Lateral ventricle

Hippocampus

Lateral ventricle

8

8

Amygdala

7

Lingula?

3

Occipital lobe

Primary olfactory cortex (piriform)

Uncus

9

4

13

Parahippocampal gyrus

Cerebellum
(Hemisphere)

-12

PLATE 6A
CR 270 mm
GW 32, Y15-60
Sagittal
Section 421

See detail of the brain core and
cerebellum in Plates 12A and B.

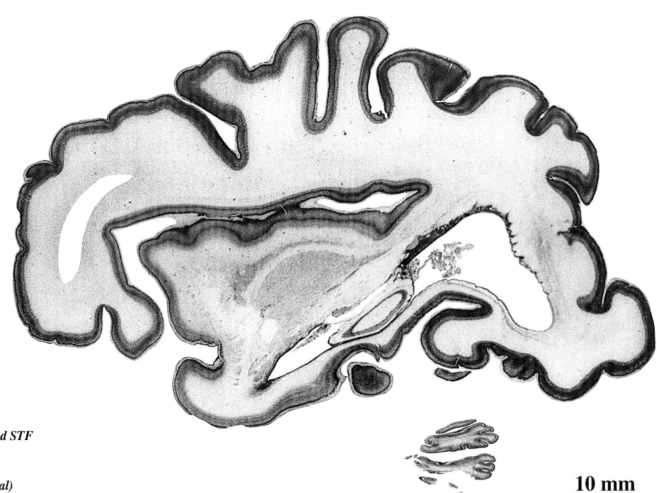

Remnants of the
germinal matrix,
migratory streams,
and transitional fields

1 *Occipital NEP, SVZ, and STF*

2 *Parahippocampal NEP, SVZ, and STF*

3 *Alvear GEP*

4 *Lateral migratory stream (cortical)*

5 *Amygdaloid G/EP*

6 *Posterior striatal NEP and SVZ*

7 *Strionuclear GEP*

8 *External germinal layer (cerebellum)*

9 *Subpial granular layer (cortical)*

10 mm

GEP - Glioepithelium
G/EP - Glioepithelium/ependyma
NEP - Neuroepithelium
STF - Stratified transitional field
SVZ - Subventricular zone

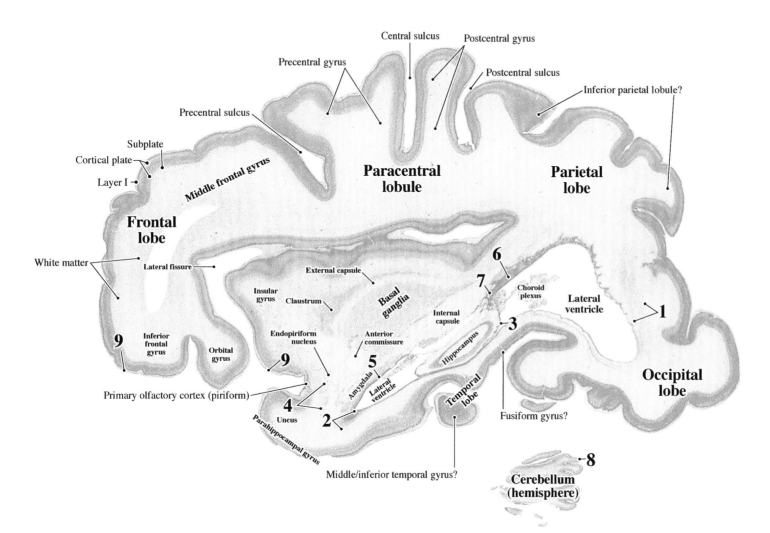

Central sulcus

Postcentral gyrus

Precentral gyrus

Postcentral sulcus

Precentral sulcus

Inferior parietal lobule?

Paracentral lobule

Parietal lobe

Subplate

Cortical plate

Middle frontal gyrus

Layer I

Frontal lobe

White matter

Lateral fissure

External capsule

6

Insular gyrus

Claustrum

Basal ganglia

7

Choroid plexus

Lateral ventricle

Internal capsule

1

9

Inferior frontal gyrus

Orbital gyrus

Endopiriform nucleus

Anterior commissure

3

Hippocampus

9

5

Amygdala

Occipital lobe

Primary olfactory cortex (piriform)

4

Lateral ventricle

Temporal lobe

2

Parahippocampal gyrus

Uncus

Fusiform gyrus?

8

Middle/inferior temporal gyrus?

Cerebellum (hemisphere)

PLATE 7A
CR 270 mm
GW 32, Y15-60
Sagittal
Section 801

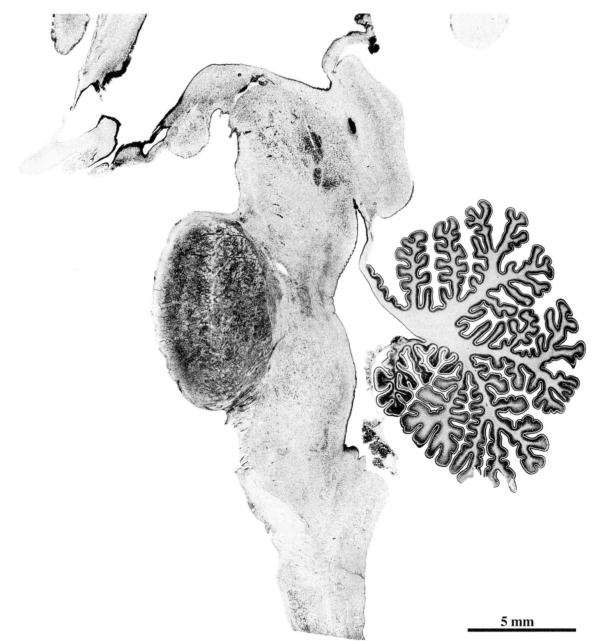

5 mm

See the entire Section 801 in Plates 1A and B.

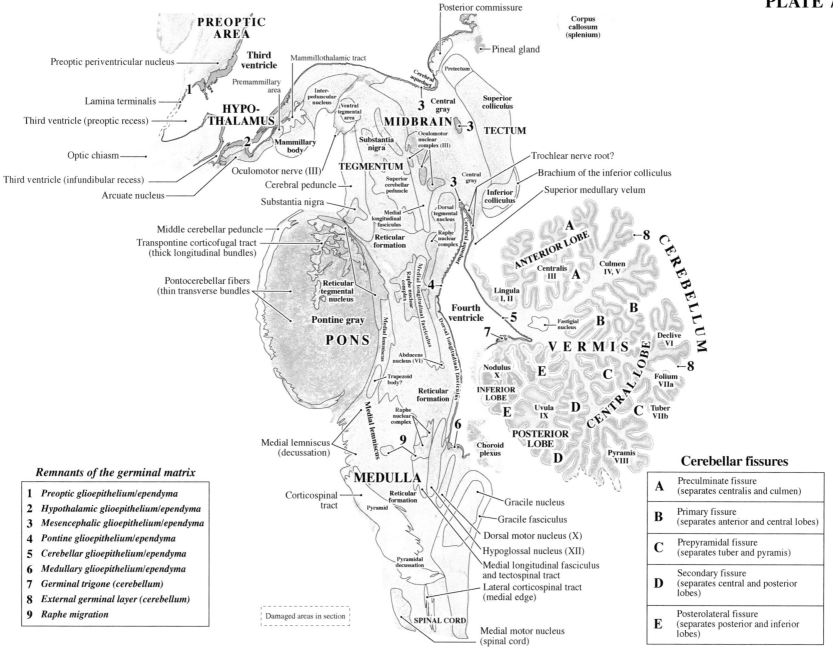

PREOPTIC AREA

Preoptic periventricular nucleus

Third ventricle

Mammillothalamic tract

Posterior commissure

Corpus callosum (splenium)

Pineal gland

Lamina terminalis

Premammillary area

Inter-peduncular nucleus

Cerebral aqueduct

Pretectum

Superior colliculus

Third ventricle (preoptic recess)

HYPO-THALAMUS

Ventral tegmental area

Central gray

MIDBRAIN

TECTUM

Optic chiasm

Mammillary body

Substantia nigra

Oculomotor nuclear complex (III)

TEGMENTUM

Trochlear nerve root?

Oculomotor nerve (III)

Cerebral peduncle

Superior cerebellar peduncle

Central gray

Brachium of the inferior colliculus

Third ventricle (infundibular recess)

Arcuate nucleus

Substantia nigra

Medial longitudinal fasciculus

Dorsal tegmental nucleus

Inferior colliculus

Superior medullary velum

Middle cerebellar peduncle

Transpontine corticofugal tract (thick longitudinal bundles)

Reticular formation

Raphe nuclear complex

ANTERIOR LOBE

A

A

CEREBELLUM

Centralis III

Culmen IV, V

Pontocerebellar fibers (thin transverse bundles)

Reticular tegmental nucleus

Raphe nuclear complex

Lingula I, II

B

Pontine gray

PONS

Medial longitudinal fasciculus

Medial lemniscus

Dorsal longitudinal fasciculus

Fourth ventricle

Fastigial nucleus

B

Declive VI

VERMIS

B

Nodulus X

E

CENTRAL LOBE

Folium VIIa

Abducens nucleus (VI)

Reticular formation

INFERIOR LOBE

E

Uvula IX

C

Tuber VIIb

Trapezoid body?

Medial lemniscus

D

POSTERIOR LOBE

Raphe nuclear complex

Medial lemniscus (decussation)

Reticular formation

Choroid plexus

Pyramis VIII

D

MEDULLA

Corticospinal tract

Reticular formation

Pyramid

Gracile nucleus

Gracile fasciculus

Dorsal motor nucleus (X)

Hypoglossal nucleus (XII)

Medial longitudinal fasciculus and tectospinal tract

Pyramidal decussation

Lateral corticospinal tract (medial edge)

SPINAL CORD

Medial motor nucleus (spinal cord)

Remnants of the germinal matrix

1 *Preoptic glioepithelium/ependyma*
2 *Hypothalamic glioepithelium/ependyma*
3 *Mesencephalic glioepithelium/ependyma*
4 *Pontine glioepithelium/ependyma*
5 *Cerebellar glioepithelium/ependyma*
6 *Medullary glioepithelium/ependyma*
7 *Germinal trigone (cerebellum)*
8 *External germinal layer (cerebellum)*
9 *Raphe migration*

Damaged areas in section

Cerebellar fissures

A	Preculminate fissure (separates centralis and culmen)
B	Primary fissure (separates anterior and central lobes)
C	Prepyramidal fissure (separates tuber and pyramis)
D	Secondary fissure (separates central and posterior lobes)
E	Posterolateral fissure (separates posterior and inferior lobes)

22

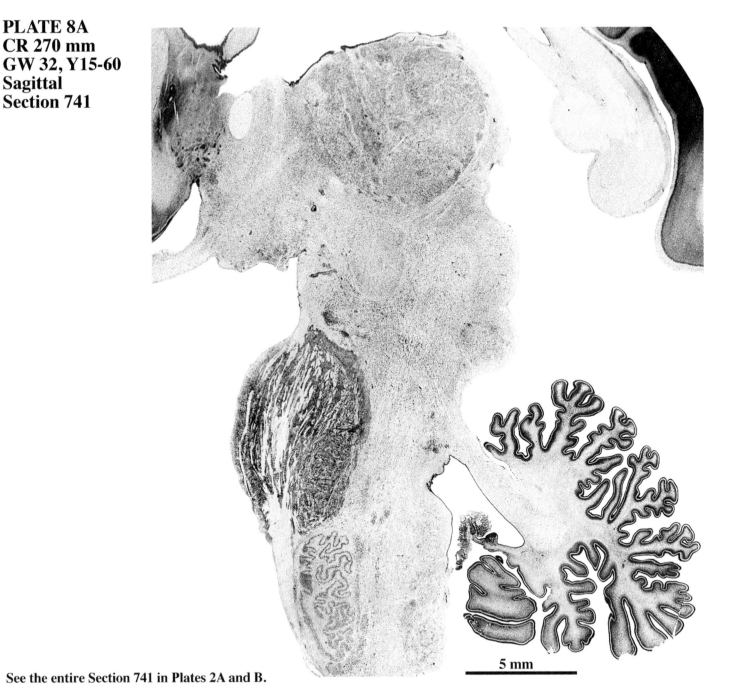

PLATE 8A
CR 270 mm
GW 32, Y15-60
Sagittal
Section 741

5 mm

See the entire Section 741 in Plates 2A and B.

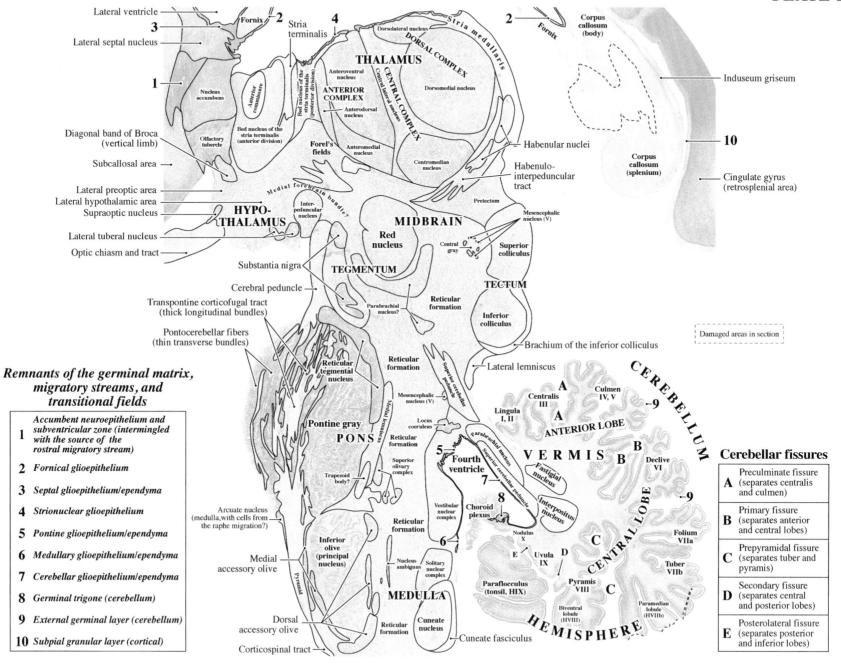

Lateral ventricle

Lateral septal nucleus

Fornix

Stria terminalis

Dorsolateral nucleus

DORSAL COMPLEX

Stria medullaris

Fornix

Corpus callosum (body)

Induseum griseum

THALAMUS

Anteroventral nucleus

ANTERIOR COMPLEX

Central lateral nucleus

CENTRAL COMPLEX

Dorsomedial nucleus

Nucleus accumbens

Anterior commissure

Bed nucleus of the stria terminalis (posterior division)

Anterodorsal nucleus

Corpus callosum (splenium)

Diagonal band of Broca (vertical limb)

Olfactory tubercle

Bed nucleus of the stria terminalis (anterior division)

Forel's fields

Anteromedial nucleus

Habenular nuclei

Centromedian nucleus

Habenulo-interpeduncular tract

Cingulate gyrus (retrosplenial area)

Subcallosal area

Lateral preoptic area

Lateral hypothalamic area

Supraoptic nucleus

Medial forebrain bundle?

Inter-peduncular nucleus

HYPO-THALAMUS

MIDBRAIN

Pretectum

Mesencephalic nucleus (V)

Lateral tuberal nucleus

Optic chiasm and tract

Red nucleus

Central gray

Superior colliculus

Substantia nigra

TEGMENTUM

Cerebral peduncle

Parabrachial nucleus?

Reticular formation

TECTUM

Transpontine corticofugal tract (thick longitudinal bundles)

Pontocerebellar fibers (thin transverse bundles)

Reticular tegmental nucleus

Reticular formation

Superior cerebellar peduncle

Inferior colliculus

Brachium of the inferior colliculus

Lateral lemniscus

CEREBELLUM

A Centralis III

Culmen IV, V

Lingula I, II

A

Mesencephalic nucleus (V)

ANTERIOR LOBE

Locus coeruleus

VERMIS

B

Declive VI

Pontine gray

PONS

Medial lemniscus

Reticular formation

Parabrachial nucleus

B

Superior olivary complex

Fourth ventricle

Superior cerebellar peduncle

Fastigial nucleus

CENTRAL LOBE

Folium VIIa

Trapezoid body?

Interpositus nucleus

Arcuate nucleus (medulla, with cells from the raphe migration?)

Reticular formation

Vestibular nuclear complex

Choroid plexus

Nodulus X

Tuber VIIb

Inferior olive (principal nucleus)

Nucleus ambiguus

Solitary nuclear complex

Uvula IX

Pyramis VIII

C

Medial accessory olive

Pyramid

MEDULLA

Parafloculus (tonsil, HIX)

Biventral lobule (HVIII)

C

Paramedian lobule (HVIIb)

Dorsal accessory olive

Reticular formation

Cuneate nucleus

HEMISPHERE

Corticospinal tract

Cuneate fasciculus

Damaged areas in section

Cerebellar fissures

A	Preculminate fissure (separates centralis and culmen)
B	Primary fissure (separates anterior and central lobes)
C	Prepyramidal fissure (separates tuber and pyramis)
D	Secondary fissure (separates central and posterior lobes)
E	Posterolateral fissure (separates posterior and inferior lobes)

24

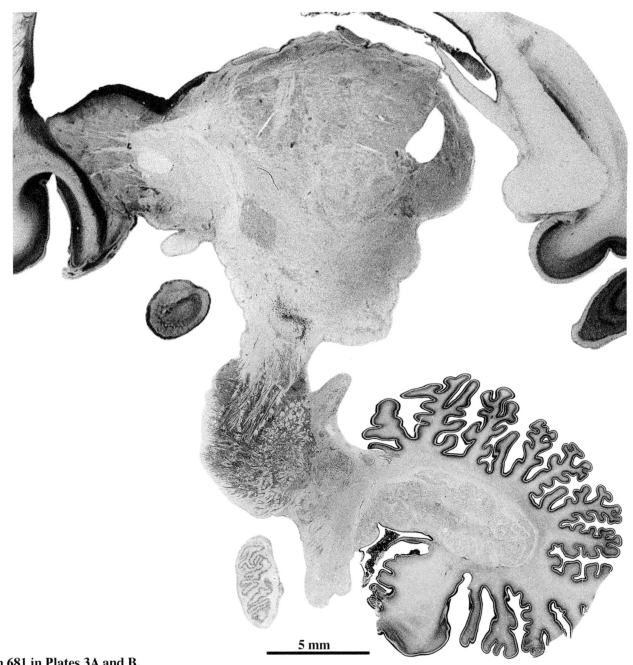

5 mm

See the entire Section 681 in Plates 3A and B.

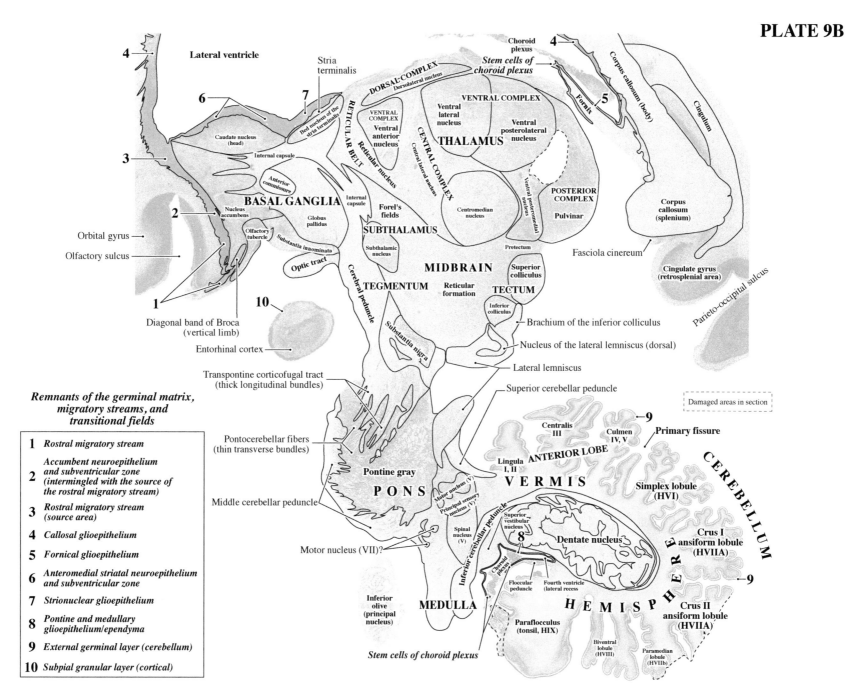

Lateral ventricle

Stria
terminalis

Choroid
plexus

*Stem cells of
choroid plexus*

Corpus callosum (body)

Cingulum

DORSAL COMPLEX
Dorsolateral nucleus

VENTRAL COMPLEX

Ventral
lateral
nucleus

Ventral
posterolateral
nucleus

THALAMUS

RETICULAR BELT

VENTRAL
COMPLEX

Ventral
anterior
nucleus

Reticular
nucleus

CENTRAL COMPLEX

Central lateral nucleus

Fornix

5

6

7

4

Caudate nucleus
(head)

Bed nucleus of the
stria terminalis

Internal capsule

3

Anterior
commissure

Internal
capsule

Forel's
fields

Ventral
posteromedial
nucleus

POSTERIOR
COMPLEX

Pulvinar

Corpus
callosum
(splenium)

2

Nucleus
accumbens

BASAL GANGLIA

Globus
pallidus

SUBTHALAMUS

Centromedian
nucleus

Pretectum

Fasciola cinereum

Orbital gyrus

Olfactory
tubercle

Substantia innominata

Subthalamic
nucleus

MIDBRAIN

Superior
colliculus

Cingulate gyrus
(retrosplenial area)

Olfactory sulcus

Optic tract

TEGMENTUM

Reticular
formation

TECTUM

Inferior
colliculus

Parieto-occipital sulcus

1

Cerebral peduncle

Substantia nigra

Brachium of the inferior colliculus

Nucleus of the lateral lemniscus (dorsal)

Lateral lemniscus

10

Diagonal band of Broca
(vertical limb)

Entorhinal cortex

Transpontine corticofugal tract
(thick longitudinal bundles)

Superior cerebellar peduncle

Damaged areas in section

*Remnants of the germinal matrix,
migratory streams, and
transitional fields*

Pontocerebellar fibers
(thin transverse bundles)

Centralis
III

Culmen
IV, V

9

Primary fissure

CEREBELLUM

Lingula
I, II

ANTERIOR LOBE

Simplex lobule
(HVI)

1 *Rostral migratory stream*

2 *Accumbent neuroepithelium
and subventricular zone
(intermingled with the source of
the rostral migratory stream)*

Middle cerebellar peduncle

Pontine gray

PONS

VERMIS

Motor nucleus (V)

Principal sensory
nucleus (V)

Crus I
ansiform lobule
(HVIIA)

3 *Rostral migratory stream
(source area)*

4 *Callosal glioepithelium*

Spinal
nucleus
(V)

Superior
vestibular
nucleus

8

Dentate nucleus

HEMISPHERE

5 *Fornical glioepithelium*

6 *Anteromedial striatal neuroepithelium
and subventricular zone*

Motor nucleus (VII)?

Inferior cerebellar peduncle

Choroid
plexus

Flocular
peduncle

Fourth ventricle
(lateral recess)

9

7 *Strionuclear glioepithelium*

8 *Pontine and medullary
glioepithelium/ependyma*

Inferior
olive
(principal
nucleus)

MEDULLA

Crus II
ansiform lobule
(HVIIA)

9 *External germinal layer (cerebellum)*

10 *Subpial granular layer (cortical)*

Paraflocculus
(tonsil, HIX)

Biventral
lobule
(HVIII)

Paramedian
lobule
(HVIIb)

Stem cells of choroid plexus

PLATE 10A
CR 270 mm
GW 32, Y15-60
Sagittal
Section 581

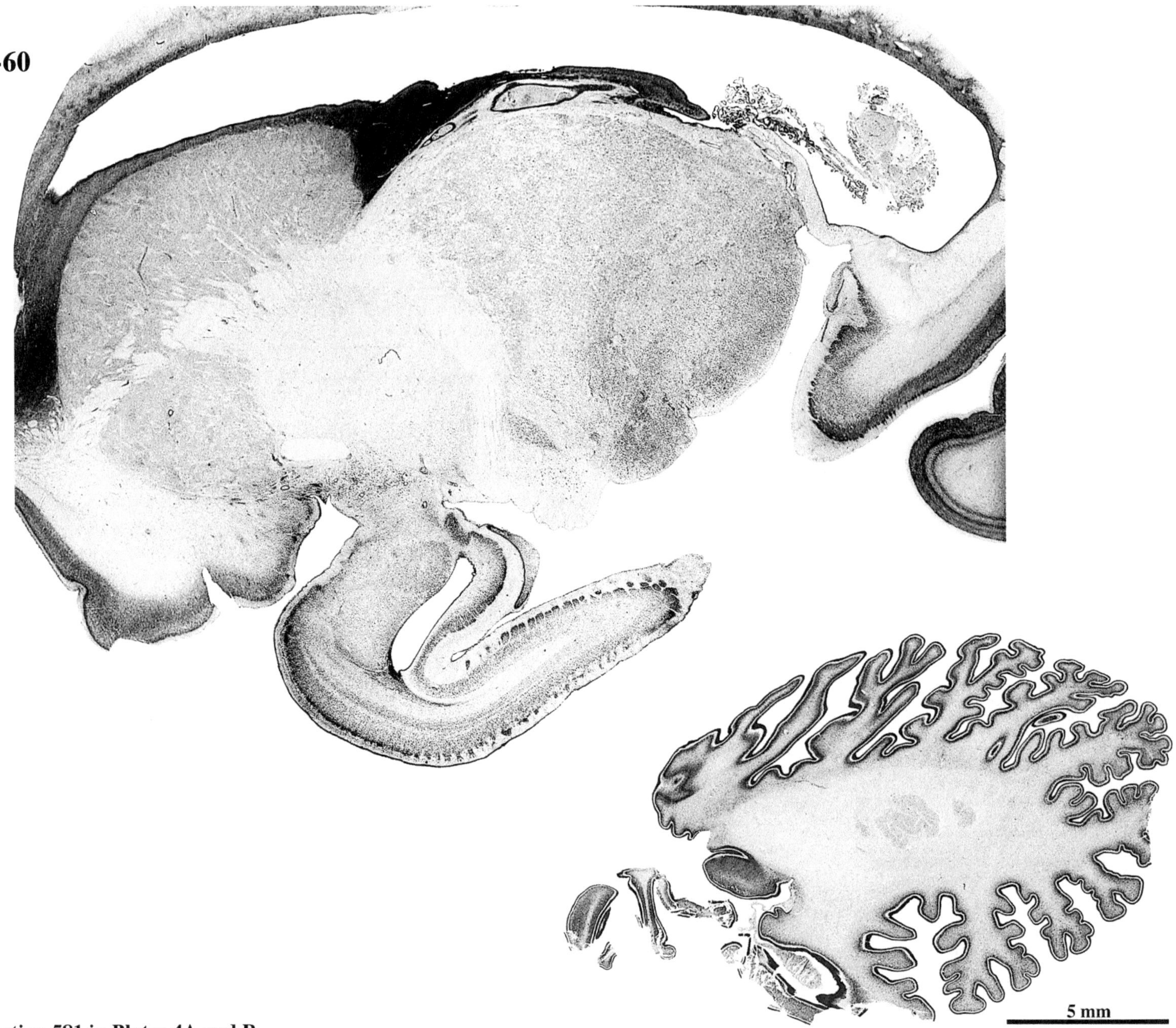

5 mm

See the entire Section 581 in Plates 4A and B.

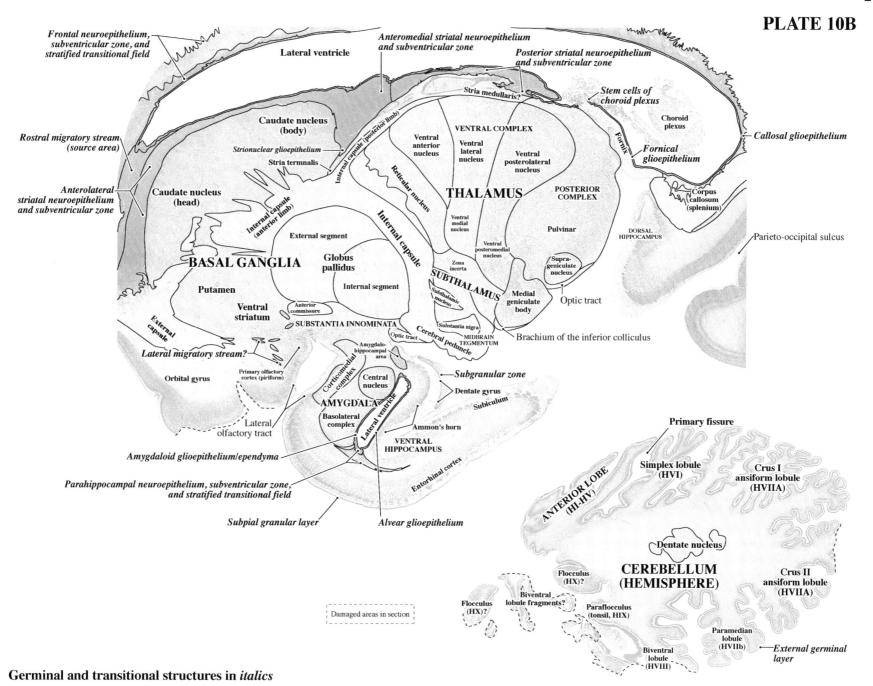

Frontal neuroepithelium, subventricular zone, and stratified transitional field

Lateral ventricle

Anteromedial striatal neuroepithelium and subventricular zone

Posterior striatal neuroepithelium and subventricular zone

Stria medullaris?

Stem cells of choroid plexus

Choroid plexus

Callosal glioepithelium

Caudate nucleus (body)

Strionuclear glioepithelium

Stria termnalis

Internal capsule (posterior limb)

VENTRAL COMPLEX

Ventral anterior nucleus

Ventral lateral nucleus

Ventral posterolateral nucleus

THALAMUS

POSTERIOR COMPLEX

Fornix

Fornical glioepithelium

Corpus callosum (splenium)

Rostral migratory stream (source area)

Caudate nucleus (head)

Reticular nucleus

Pulvinar

DORSAL HIPPOCAMPUS

Parieto-occipital sulcus

Anterolateral striatal neuroepithelium and subventricular zone

Internal capsule (anterior limb)

Internal capsule

External segment

BASAL GANGLIA

Globus pallidus

Ventral medial nucleus

Ventral posteromedial nucleus

Zona incerta

Supra-geniculate nucleus

Putamen

Internal segment

SUBTHALAMUS

Subthalamic nucleus

Medial geniculate body

Optic tract

Ventral striatum

Anterior commissure

SUBSTANTIA INNOMINATA

Substantia nigra

Cerebral peduncle

MIDBRAIN TEGMENTUM

Brachium of the inferior colliculus

External capsule

Lateral migratory stream?

Optic tract

Orbital gyrus

Primary olfactory cortex (piriform)

Amygdalo-hippocampal area

Subgranular zone

Corticomedial complex

Central nucleus

Dentate gyrus

Subiculum

AMYGDALA

Basolateral complex

Ammon's horn

Lateral olfactory tract

Lateral ventricle

VENTRAL HIPPOCAMPUS

Primary fissure

Amygdaloid glioepithelium/ependyma

Entorhinal cortex

ANTERIOR LOBE (HI-HV)

Simplex lobule (HVI)

Crus I ansiform lobule (HVIIA)

Parahippocampal neuroepithelium, subventricular zone, and stratified transitional field

Dentate nucleus

CEREBELLUM (HEMISPHERE)

Crus II ansiform lobule (HVIIA)

Subpial granular layer

Alvear glioepithelium

Flocculus (HX)?

Flocculus (HX)?

Biventral lobule fragments?

Paraflocculus (tonsil, HIX)

Paramedian lobule (HVIIb)

External germinal layer

Damaged areas in section

Biventral lobule (HVIII)

Germinal and transitional structures in *italics*

PLATE 11A
CR 270 mm
GW 32, Y15-60
Sagittal
Section 481

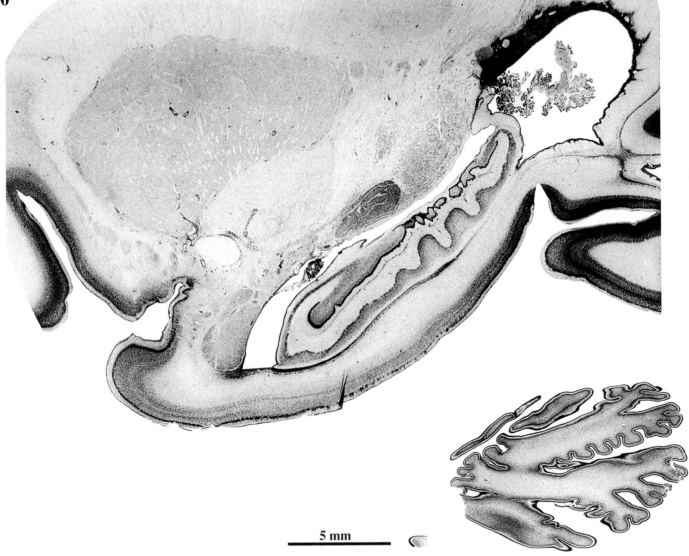

5 mm

See the entire Section 481 in Plates 5A and B.

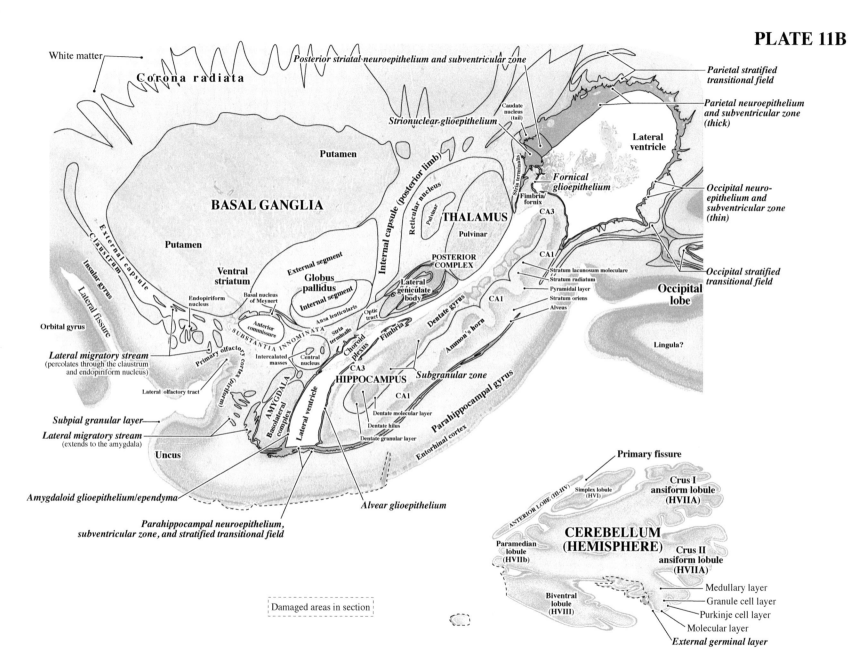

White matter

Corona radiata

Posterior striatal neuroepithelium and subventricular zone

Parietal stratified transitional field

Strionuclear-glioepithelium

Caudate nucleus (tail)

Parietal neuroepithelium and subventricular zone (thick)

Putamen

Stria terminalis

Fornical glioepithelium

Lateral ventricle

BASAL GANGLIA

Internal capsule (posterior limb)

Reticular nucleus

Pulvinar

THALAMUS

CA3

Fimbria/fornix

Occipital neuro-epithelium and subventricular zone (thin)

Putamen

External capsule

Claustrum

Insular gyrus

Lateral fissure

Endopiriform nucleus

Basal nucleus of Meynert

Ventral striatum

External segment

Globus pallidus

Internal segment

Ansa lenticularis

Optic tract

POSTERIOR COMPLEX

Pulvinar

Lateral geniculate body

CA1

CA1

Stratum lacunosum moleculare

Stratum radiatum

Pyramidal layer

Stratum oriens

Alveus

Dentate gyrus

CA1

Occipital lobe

Occipital stratified transitional field

Orbital gyrus

Anterior commissure

SUBSTANTIA INNOMINATA

Stria terminalis

Choroid plexus

Fimbria

Ammon's horn

Lingula?

Lateral migratory stream
(percolates through the claustrum and endopiriform nucleus)

Primary olfactory cortex (piriform)

Intercalated masses

Central nucleus

CA3

HIPPOCAMPUS

Subgranular zone

Lateral olfactory tract

CA1

Parahippocampal gyrus

Subpial granular layer

AMYGDALA

Basolateral complex

Lateral ventricle

Dentate molecular layer

Dentate hilus

Dentate granular layer

Lateral migratory stream
(extends to the amygdala)

Uncus

Entorhinal cortex

Amygdaloid glioepithelium/ependyma

Alvear glioepithelium

Parahippocampal neuroepithelium, subventricular zone, and stratified transitional field

Primary fissure

ANTERIOR LOBE (HI-HIV)

Simplex lobule (HVI)

Crus I ansiform lobule (HVIIA)

Paramedian lobule (HVIIb)

CEREBELLUM (HEMISPHERE)

Crus II ansiform lobule (HVIIA)

Damaged areas in section

Biventral lobule (HVIII)

Medullary layer

Granule cell layer

Purkinje cell layer

Molecular layer

External germinal layer

Germinal and transitional structures in *italics*

PLATE 12A, CR 270 mm, GW 32, Y15-60
Sagittal, Section 421

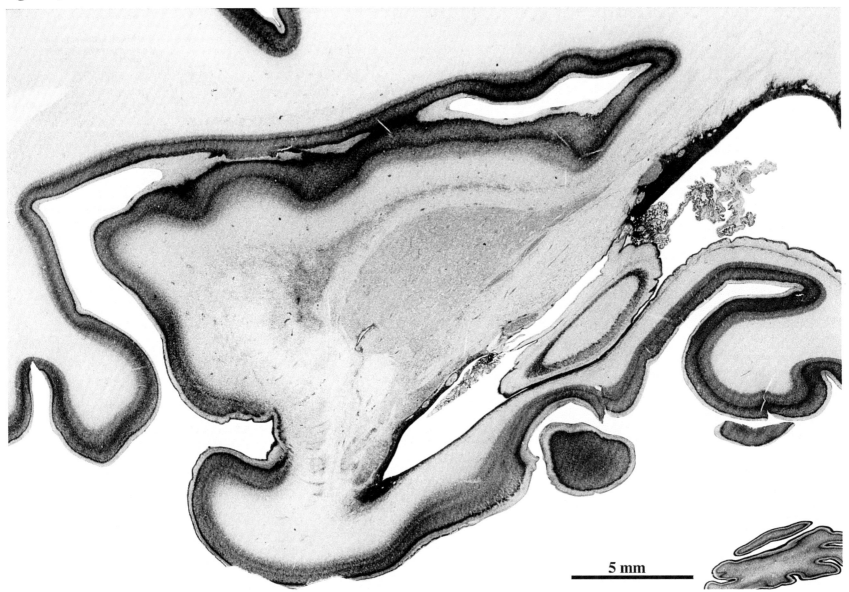

5 mm

See the entire Section 421 in Plates 6A and B.

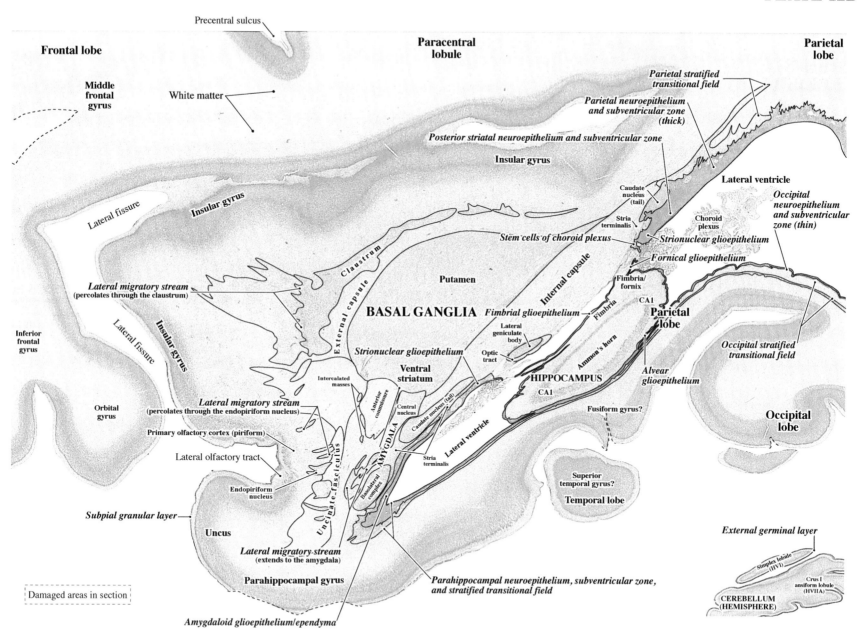

Precentral sulcus

Paracentral lobule

Frontal lobe

Parietal lobe

Middle frontal gyrus

White matter

Parietal stratified transitional field

Parietal neuroepithelium and subventricular zone (thick)

Posterior striatal neuroepithelium and subventricular zone

Insular gyrus

Lateral ventricle

Insular gyrus

Caudate nucleus (tail)

Occipital neuroepithelium and subventricular zone (thin)

Lateral fissure

Choroid plexus

Stria terminalis

Stem cells of choroid plexus

Strionuclear glioepithelium

Claustrum

Fornical glioepithelium

Lateral migratory stream (percolates through the claustrum)

Putamen

Fimbria/fornix

BASAL GANGLIA

Fimbrial glioepithelium

Fimbria

CA1

External capsule

Internal capsule

Parietal lobe

Inferior frontal gyrus

Lateral fissure

Insular gyrus

Lateral geniculate body

Strionuclear glioepithelium

Ammon's horn

Occipital stratified transitional field

Optic tract

Alvear glioepithelium

Ventral striatum

HIPPOCAMPUS

CA1

Orbital gyrus

Intercalated masses

Anterior commissure

Central nucleus

Caudate nucleus (tail)

Fusiform gyrus?

Occipital lobe

Lateral migratory stream (percolates through the endopiriform nucleus)

Lateral ventricle

AMYGDALA

Stria terminalis

Primary olfactory cortex (piriform)

Lateral olfactory tract

Endopiriform nucleus

Uncinate fasciculus

Basolateral complex

Superior temporal gyrus?

Temporal lobe

Subpial granular layer

External germinal layer

Uncus

Lateral migratory stream (extends to the amygdala)

Parahippocampal gyrus

Parahippocampal neuroepithelium, subventricular zone, and stratified transitional field

Simplex lobile (HVI)

Crus I ansiform lobule (HVIIA)

CEREBELLUM (HEMISPHERE)

Damaged areas in section

Amygdaloid glioepithelium/ependyma

Germinal and transitional structures in *italics*

32

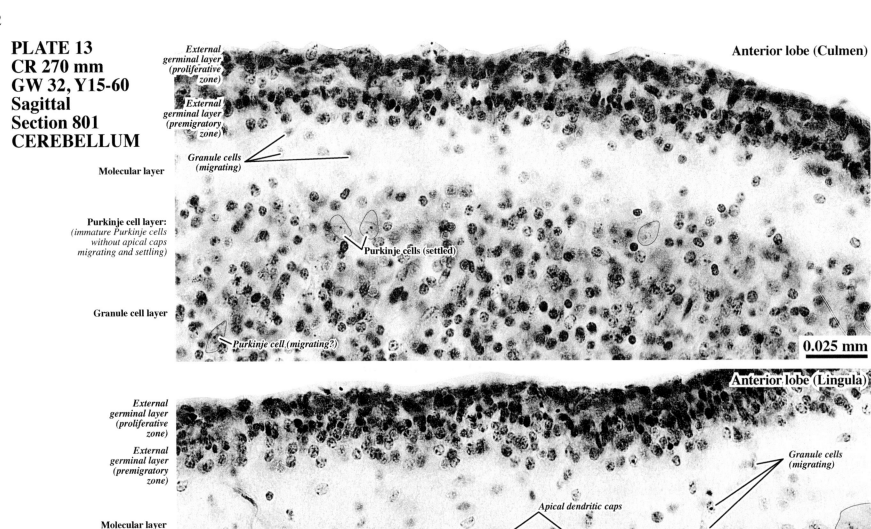

PLATE 13
CR 270 mm
GW 32, Y15-60
Sagittal
Section 801
CEREBELLUM

Anterior lobe (Culmen)

External germinal layer (proliferative zone)

External germinal layer (premigratory zone)

Granule cells (migrating)

Molecular layer

Purkinje cell layer:
(immature Purkinje cells without apical caps migrating and settling)

Purkinje cells (settled)

Granule cell layer

Purkinje cell (migrating?)

0.025 mm

Anterior lobe (Lingula)

External germinal layer (proliferative zone)

External germinal layer (premigratory zone)

Granule cells (migrating)

Molecular layer

Apical dendritic caps

Purkinje cell layer:
(Maturing Purkinje cells with prominent apical dendritic caps settled in a monolayer)

Granule cell layer

0.025 mm

See the entire Section 801 in Plates 1A and B.

A medium-magnification view of the cerebellum is in Plates 7A and B.

Central lobe (Folium)

**PLATE 14
CR 270 mm
GW 32, Y15-60
Sagittal
Section 801
CEREBELLUM**

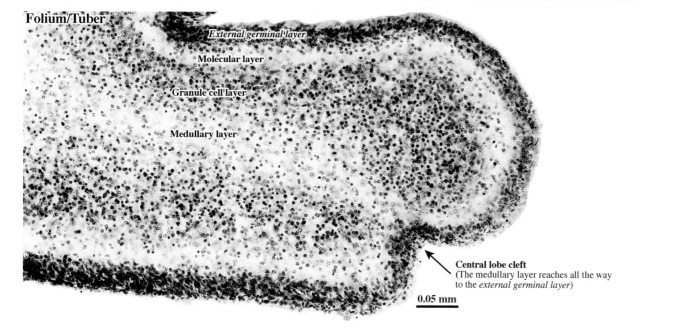

External germinal layer (proliferative zone)

External germinal layer (premigratory zone)

Granule cells (migrating)

Settling small Purkinje cells

Molecular layer

Purkinje cell layer:
(scattered *immature Purkinje cells migrating and settling*)

Migrating small Purkinje cells

Granule cell layer

0.025 mm

Folium/Tuber

External germinal layer

Molecular layer

Granule cell layer

Medullary layer

Central lobe cleft
(The medullary layer reaches all the way to the *external germinal layer*)

0.05 mm

See the entire Section 801 in Plates 1A and B.

A medium-magnification view of the cerebellum is in Plates 7A and B.

**PLATE 15
CR 270 mm
GW 32, Y15-60
Sagittal
Section 801
CEREBELLUM**

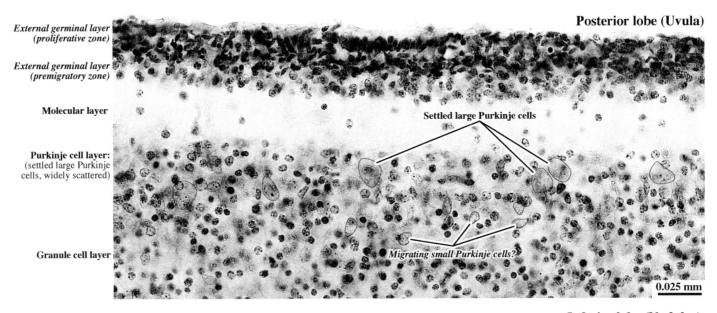

Posterior lobe (Uvula)

*External germinal layer
(proliferative zone)*

*External germinal layer
(premigratory zone)*

Molecular layer

Settled large Purkinje cells

Purkinje cell layer:
(settled large Purkinje
cells, widely scattered)

Granule cell layer

Migrating small Purkinje cells?

0.025 mm

Inferior lobe (Nodulus)

*External germinal layer
(proliferative zone)*

*External germinal layer
(premigratory zone)*

*Granule cells
(migrating)*

Molecular layer

Purkinje cell layer:
(Purkinje cells with
*basal cytoplasmic
accumulations* settled
in a monolayer)

Basal cytoplasmic accumulations

Granule cell layer

0.025 mm

**See the entire Section 801
in Plates 1A and B.**

**A medium-magnification
view of the cerebellum is in
Plates 7A and B.**

PLATE 16
CR 270 mm
GW 32, Y15-60
Sagittal
Section 801
CEREBELLUM

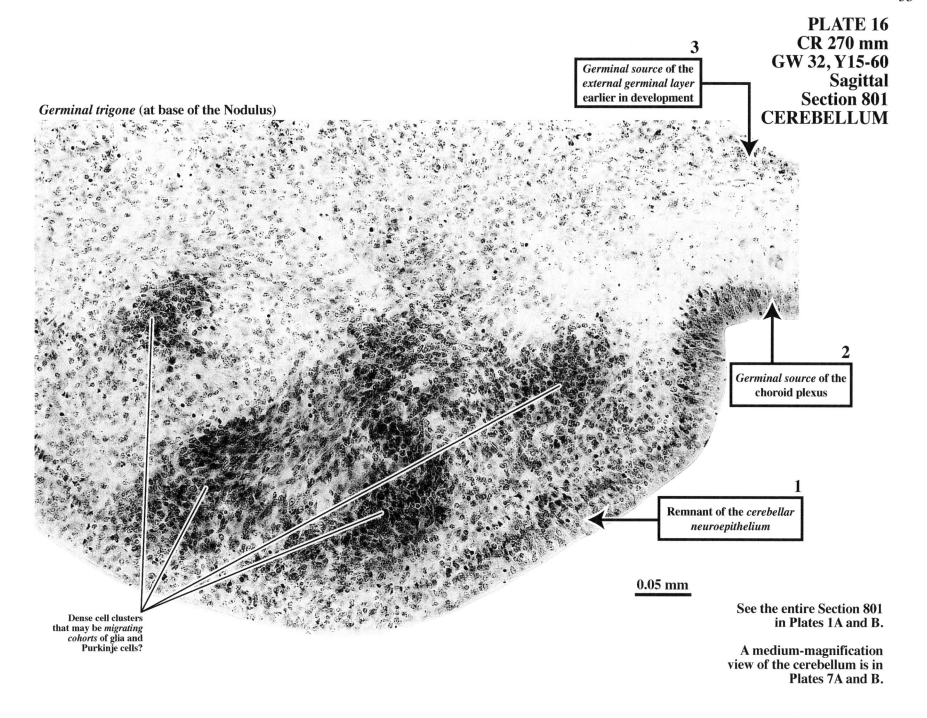

3

Germinal source of the *external germinal layer* earlier in development

Germinal trigone (at base of the Nodulus)

2

Germinal source of the choroid plexus

1

Remnant of the *cerebellar neuroepithelium*

Dense cell clusters that may be *migrating cohorts* of glia and Purkinje cells?

0.05 mm

See the entire Section 801 in Plates 1A and B.

A medium-magnification view of the cerebellum is in Plates 7A and B.

PART III: Y14-59
CR 260 mm (GW 30)
Frontal

This specimen is case number W-14-59 (Perinatal RPSL) in the Yakovlev Collection. A female fetus was prematurely stillborn after intrauterine asphyxia. Autopsy notes include a subdural hemorrhage, and there is a hemorrhage in the striatal germinal matrix on the left side of many sections. However, the remainder of the brain appears normal and is classified as a Normative Control in the Yakovlev Collection (Haleem, 1990). It was cut in the coronal plane in 35-μm and 15-μm thick sections. Since there is no available photograph of this brain before it was embedded and cut, the photograph of the lateral view of a GW 30 brain that Larroche published in 1967 (**Figure 4**) is used.

The approximate cutting plane of this brain is indicated in **Figure 5** (facing page) with lines superimposed on the GW 30 brain from the Larroche (1967) series. This brain is cut perpendicular to the longitudinal axis of the cerebral hemispheres between the frontal and occipital poles and is remarkably even in the medial/lateral plane (both temporal poles appear in **Section 621** (**Plate 19**). The sections chosen for illustration are spaced closer together to show small structures in the diencephalon, midbrain, pons, and medulla; spaced farther apart when only large brain structures, such as the cerebral cortex, basal ganglia, and cerebellum are present. Photographs of 20 low-magnification Nissl-stained sections are shown in **Plates 17-36**. Different areas of the cerebral cortex are shown at very high magnification in **Plates 37-40**. **Plates 41-55** show the brain core and cerebellum at high magnification.

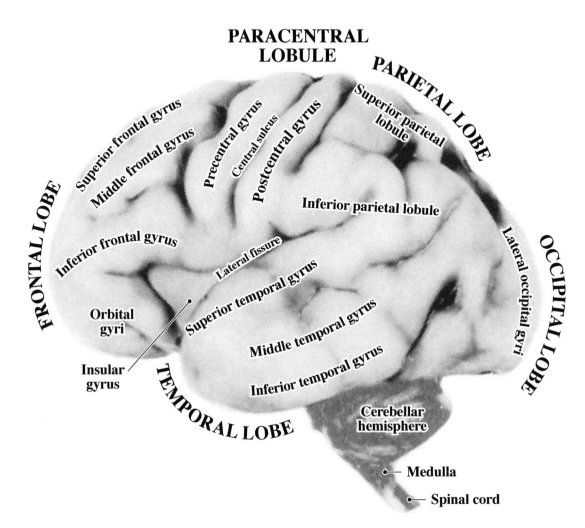

Figure 4. Lateral view of a GW 30 brain with major structures in the cerebral hemispheres labeled. This is the same as **Figure 6** repeated here for convenience. (From the photographic series of J. C. Larroche (1967) Maturation morphologique du système nerveux central: ses rapports avec le développement pondéral du foetus et son age gestationnel. In: *Regional Development of the Brain in Early Life*, A. Minkowski (ed.), London: Blackwell, page 248.)

GW30 FRONTAL SECTION PLANES
SECTION NUMBER

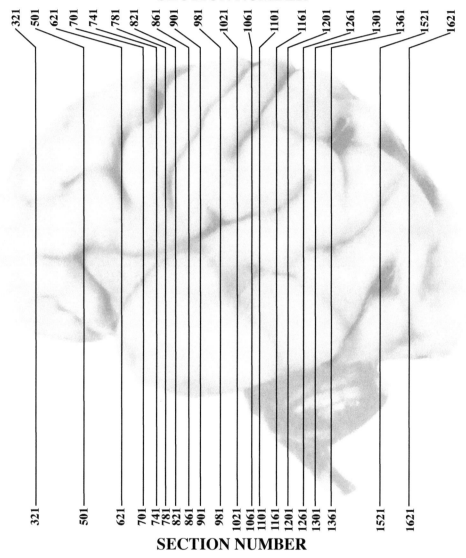

321 501 621 701 741 781 821 861 901 981 1021 1061 1101 1161 1201 1261 1301 1361 1521 1621

321 501 621 701 741 781 821 861 901 981 1021 1061 1101 1161 1201 1261 1301 1361 1521 1621

SECTION NUMBER

Figure 5. Lateral view of the same GW30 brain shown in **Figure 4** with the approximate locations and cutting angle of the sections of Y14-59. (From the photographic series of J. C. Larroche (1967) Maturation morphologique du système nerveux central: ses rapports avec le développement pondéral du foetus et son age gestationnel. In: *Regional Development of the Brain in Early Life*, A. Minkowski (ed.), London: Blackwell, page 248.)

Immature structures are in the cerebral cortex and cerebellum. A densely staining *neuroepithelium/subventricular zone* is present and presumably generating neocortical interneurons in all lobes of the cerebral cortex. Remnants of migrating and sojourning neurons and/or glia are visible in the cortical *stratified transitional fields*. Many neurons, glia, and their mitotic precursor cells are still migrating through the olfactory peduncle toward the olfactory bulb in the *rostral migratory stream*. Within the lateral parts of the cerebral cortex, neurons and glia are in the *lateral migratory stream* that percolates through the claustrum, endopiriform nucleus, external capsule, and uncinate fasciculus. These cells appear to be heading toward the insular cortex, primary olfactory cortex, temporal cortex, and basolateral parts of the amygdaloid complex. In the basal ganglia, there is a large *neuroepithelium/subventricular zone* overlying the striatum and nucleus accumbens where neurons are being generated. Another region of active neurogenesis in the telencephalon is the *subgranular zone* in the hilus of the dentate gyrus that is the source of granule cells. Other structures in the telencephalon, such as the septum, fornix, and Ammon's horn have only a thin layer at the ventricle, and these are presumed to be generating glia, cells of the choroid plexus, and the ependymal lining of the ventricle.

Most of the structures in the diencephalon appear to be settled and are maturing, and the third ventricle is lined by a densely staining *glioepithelium/ependyma*. A convoluted *glioepithelium/ependyma* lines the cerebral aqueduct in the midbrain that continues into the anterior fourth ventricle. A smooth *glioepithelium/ependyma* lines the fourth ventricle through the posterior pons. A slightly convoluted *glioepithelium/ependyma* lines the floor of the fourth ventricle near the midline in the medulla. The *external germinal layer* is prominent over the entire surface of the cerebellar cortex and is actively producing basket, stellate, and granule cells. The *germinal trigone* is at the base of the nodulus and along the floccular peduncle; choroid plexus cells and glia may still be originating here.

38

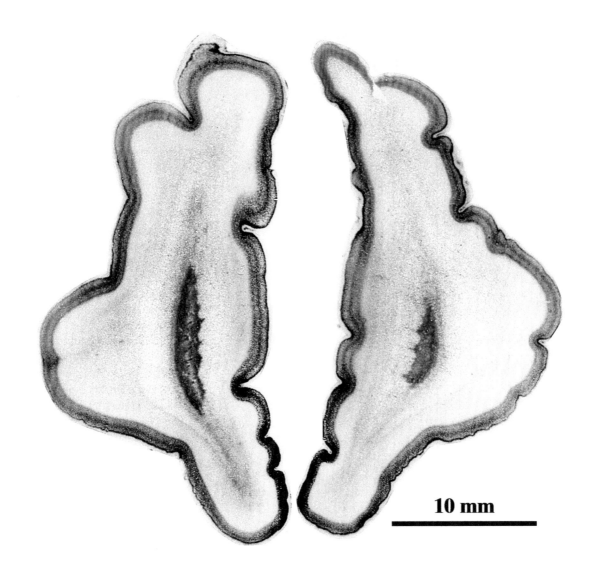

10 mm

Remnants of the germinal matrix,
migratory streams, and transitional fields

1 *Frontal NEP and SVZ*

2 *Frontal STF*

NEP - Neuroepithelium
STF - Stratified transitional field
SVZ - Subventricular zone

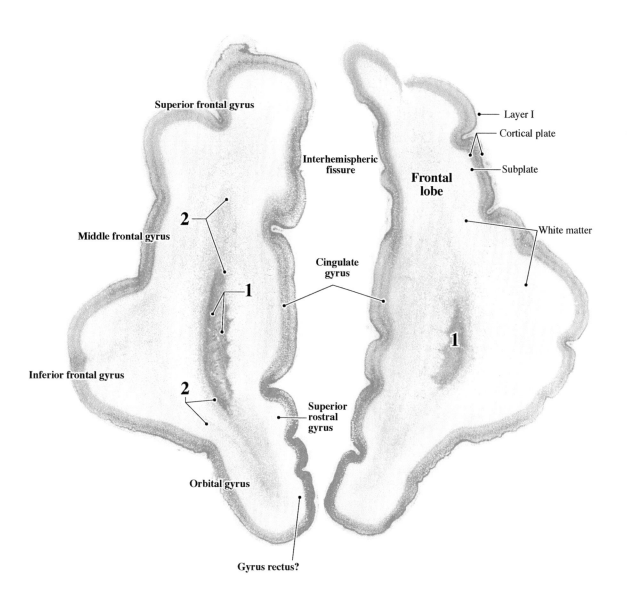

Superior frontal gyrus

Interhemispheric fissure

Layer I

Cortical plate

Frontal lobe

Subplate

2

Middle frontal gyrus

White matter

Cingulate gyrus

1

Inferior frontal gyrus

1

2

Superior rostral gyrus

Orbital gyrus

Gyrus rectus?

PLATE 18A
CR 260 mm
GW 30, Y14-59
Frontal
Section 501

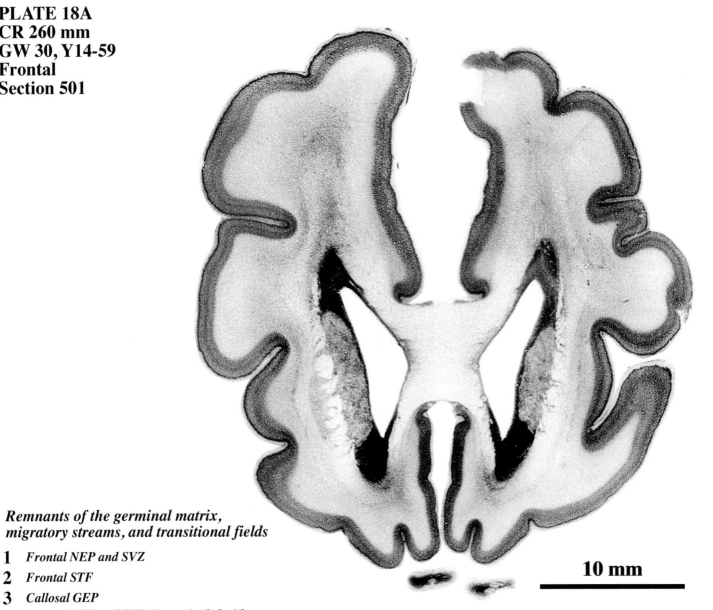

Remnants of the germinal matrix,
migratory streams, and transitional fields

1   *Frontal NEP and SVZ*

2   *Frontal STF*

3   *Callosal GEP*

4   *Frontal NEP and SVZ (intermingled with*
    the source of the rostral migratory stream)

5   *Rostral migratory stream*

6   *Anterolateral striatal NEP and SVZ*

GEP - Glioepithelium
NEP - Neuroepithelium
STF - Stratified transitional field
SVZ - Subventricular zone

10 mm

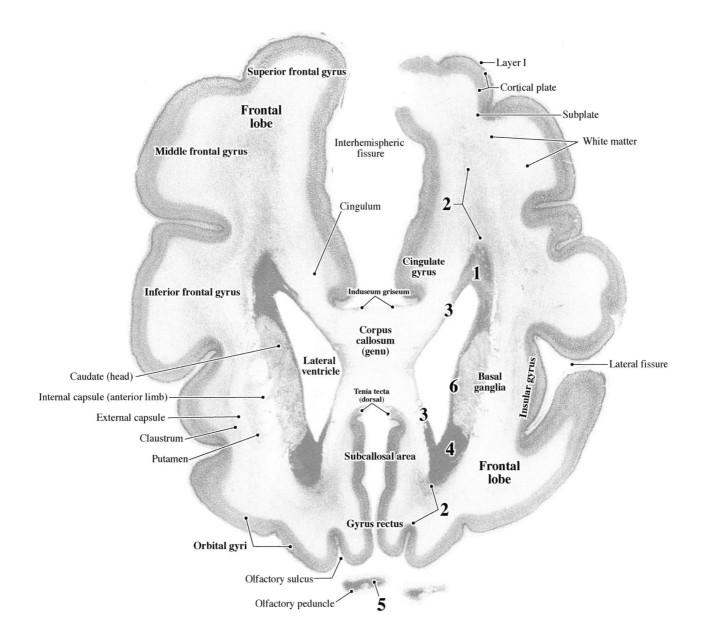

Superior frontal gyrus

Frontal
lobe

Middle frontal gyrus

Interhemispheric
fissure

Layer I

Cortical plate

Subplate

White matter

Cingulum

2

Cingulate
gyrus

1

Inferior frontal gyrus

Induseum griseum

3

Corpus
callosum
(genu)

Lateral
ventricle

Caudate (head)

Basal
ganglia

6

Lateral fissure

Internal capsule (anterior limb)

Tenia tecta
(dorsal)

3

Insular gyrus

External capsule

Claustrum

4

Putamen

Subcallosal area

Frontal
lobe

Gyrus rectus

2

Orbital gyri

Olfactory sulcus

Olfactory peduncle

5

PLATE 19A
CR 260 mm
GW 30, Y14-59
Frontal
Section 621

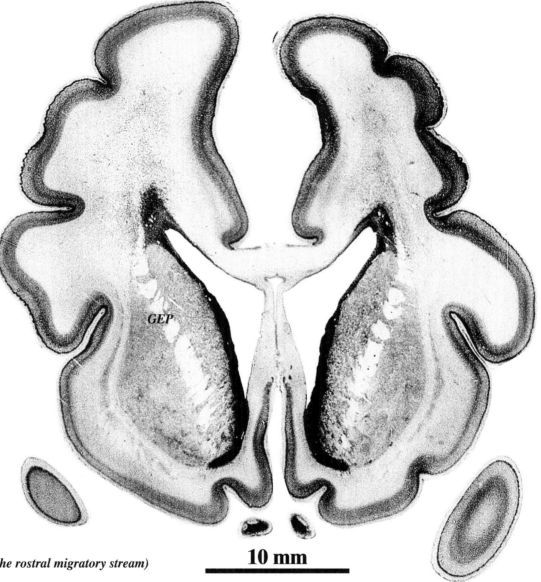

GEP

10 mm

Remnants of the germinal matrix,
migratory streams, and transitional fields

1   *Frontal NEP and SVZ*

2   *Frontal STF*

3   *Callosal GEP*

4   *Callosal sling*

5   *Fornical GEP*

6   *Rostral migratory stream*

7   *Accumbent NEP and SVZ (intermingled with the rostral migratory stream)*

8   *Anteromedial striatal NEP and SVZ*

9   *Anterolateral striatal NEP and SVZ*

10   *Lateral migratory stream (cortical)*

11   *Subpial granular layer (cortical)*

GEP - Glioepithelium
NEP - Neuroepithelium
STF - Stratified transitional field
SVZ - Subventricular zone

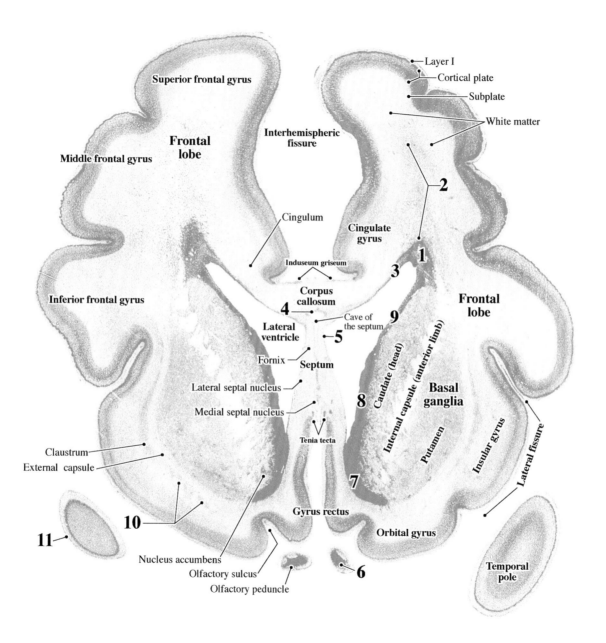

Superior frontal gyrus

Layer I

Cortical plate

Subplate

White matter

Interhemispheric
fissure

**Frontal
lobe**

Middle frontal gyrus

Cingulum

Cingulate
gyrus

2

Induseum griseum

1

3

**Corpus
callosum**

Inferior frontal gyrus

**Frontal
lobe**

4

Cave of
the septum

9

**Lateral
ventricle**

5

Fornix

Septum

Caudate (head)

Internal capsule (anterior limb)

**Basal
ganglia**

Lateral septal nucleus

8

Medial septal nucleus

Putamen

Insular gyrus

Lateral fissure

Tenia tecta

Claustrum

7

External capsule

10

Gyrus rectus

11

Orbital gyrus

Nucleus accumbens

Olfactory sulcus

6

**Temporal
pole**

Olfactory peduncle

PLATE 20A
CR 260 mm
GW 30, Y14-59
Frontal
Section 701

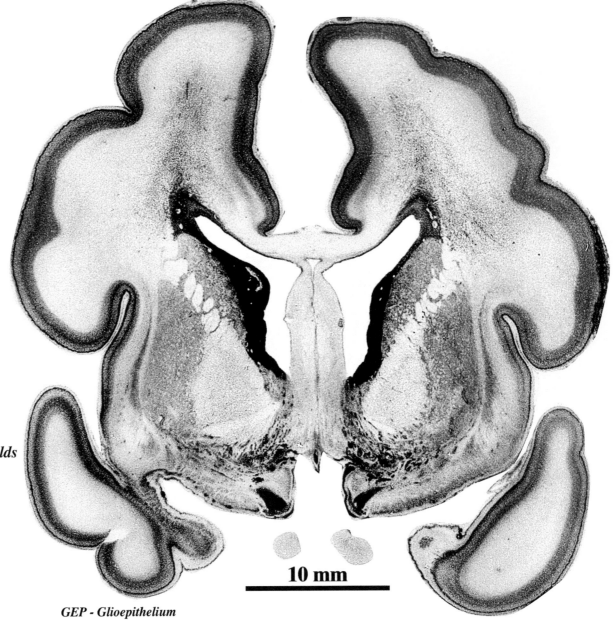

10 mm

Remnants of the germinal matrix,
migratory streams, and transitional fields

1 *Frontal NEP and SVZ*
2 *Frontal stratified transitional field*
3 *Callosal GEP*
4 *Callosal sling*
5 *Fornical GEP*
6 *Rostral migratory stream*
7 *Accumbent NEP and SVZ*
8 *Anteromedial striatal NEP and SVZ*
9 *Anterolateral striatal NEP and SVZ*
10 *Lateral migratory stream (cortical)*
11 *Subpial granular layer (cortical)*

GEP - Glioepithelium
NEP - Neuroepithelium
STF - Stratified transitional field
SVZ - Subventricular zone

See detail of the brain core
in Plates 41A and B.

PLATE 20B

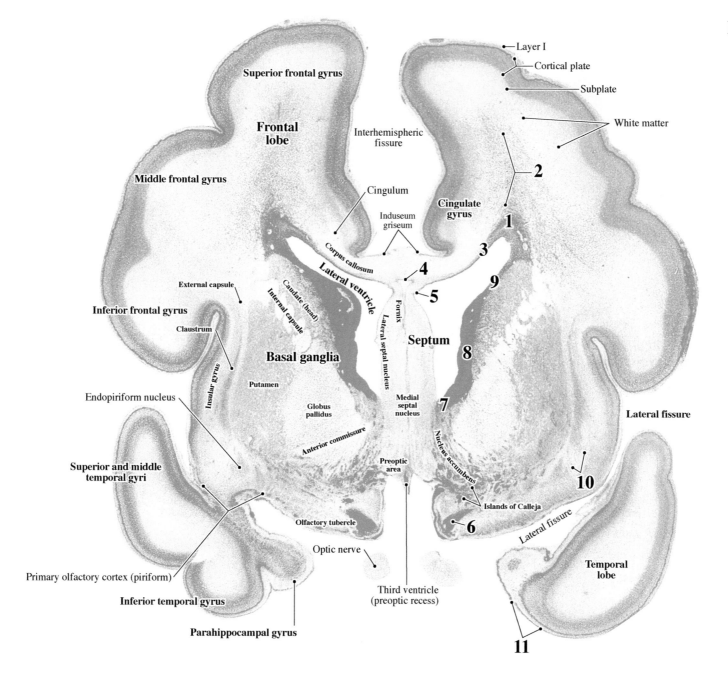

Superior frontal gyrus

Frontal lobe

Middle frontal gyrus

Interhemispheric fissure

Layer I

Cortical plate

Subplate

White matter

2

Cingulum

Induseum griseum

Cingulate gyrus

1

Corpus callosum

3

4

9

External capsule

Caudate (head)

Internal capsule

Lateral ventricle

5

Inferior frontal gyrus

Claustrum

Fornix

Lateral septal nucleus

Septum

Insular gyrus

Basal ganglia

8

Endopiriform nucleus

Putamen

Globus pallidus

Medial septal nucleus

7

Lateral fissure

Superior and middle temporal gyri

Anterior commissure

Nucleus accumbens

10

Preoptic area

Islands of Calleja

Olfactory tubercle

6

Optic nerve

Lateral fissure

Primary olfactory cortex (piriform)

Temporal lobe

Inferior temporal gyrus

Third ventricle (preoptic recess)

Parahippocampal gyrus

11

PLATE 21A
CR 260 mm
GW 30, Y14-59
Frontal
Section 741

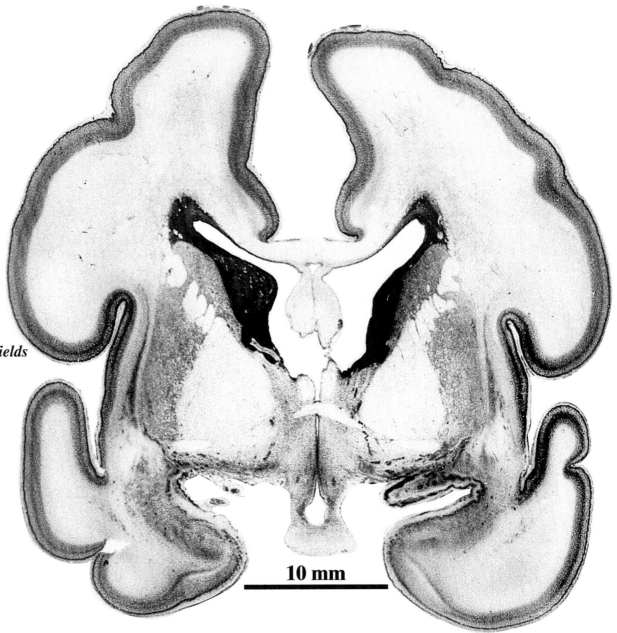

Remnants of the germinal matrix,
migratory streams, and transitional fields

 1 *Frontal NEP and SVZ*

 2 *Frontal STF*

 3 *Callosal GEP*

 4 *Callosal sling*

 5 *Fornical GEP*

 6 *Strionuclear GEP*

 7 *Anteromedial striatal NEP and SVZ*

 8 *Anterolateral striatal NEP and SVZ*

 9 *Lateral migratory stream (cortical)*

10 *Subpial granular layer (cortical)*

GEP - Glioepithelium
NEP - Neuroepithelium
STF - Stratified transitional field
SVZ - Subventricular zone

10 mm

See detail of the brain core
in Plates 42A and B.

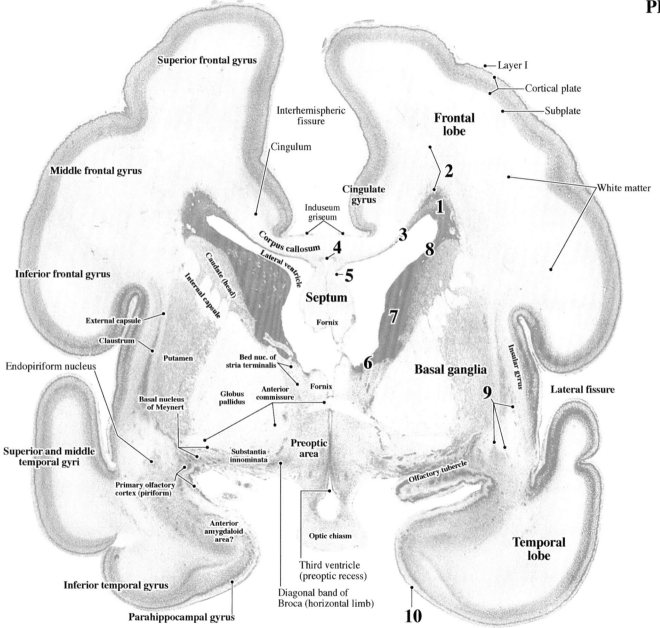

Superior frontal gyrus

Interhemispheric fissure

Layer I

Cortical plate

Subplate

Frontal lobe

Cingulum

Middle frontal gyrus

2

White matter

Cingulate gyrus

Induseum griseum

1

Corpus callosum

3

4

8

Lateral ventricle

Caudate (head)

5

Inferior frontal gyrus

Internal capsule

Septum

Fornix

7

External capsule

Claustrum

Putamen

6

Basal ganglia

Endopiriform nucleus

Bed nuc. of stria terminalis

Fornix

Insular gyrus

Lateral fissure

Basal nucleus of Meynert

Globus pallidus

Anterior commissure

9

Superior and middle temporal gyri

Substantia innominata

Preoptic area

Primary olfactory cortex (piriform)

Olfactory tubercle

Anterior amygdaloid area?

Optic chiasm

Temporal lobe

Inferior temporal gyrus

Third ventricle (preoptic recess)

Diagonal band of Broca (horizontal limb)

10

Parahippocampal gyrus

PLATE 22A
CR 260 mm
GW 30, Y14-59
Frontal
Section 781

Remnants of the germinal matrix,
migratory streams, and
transitional fields

1   *Frontal NEP and SVZ*

2   *Frontal STF*

3   *Callosal GEP*

4   *Callosal sling*

5   *Fornical GEP*

6   *Strionuclear GEP*

7   *Anteromedial striatal NEP and SVZ*

8   *Anterolateral striatal NEP and SVZ*

9   *Lateral migratory stream (cortical)*

10 *Parahippocampal STF (intermingled*
    with the amygdaloid G/EP)

GEP - Glioepithelium
G/EP - Glioepithelium/ependyma
NEP - Neuroepithelium
STF - Stratified transitional field
SVZ - Subventricular zone

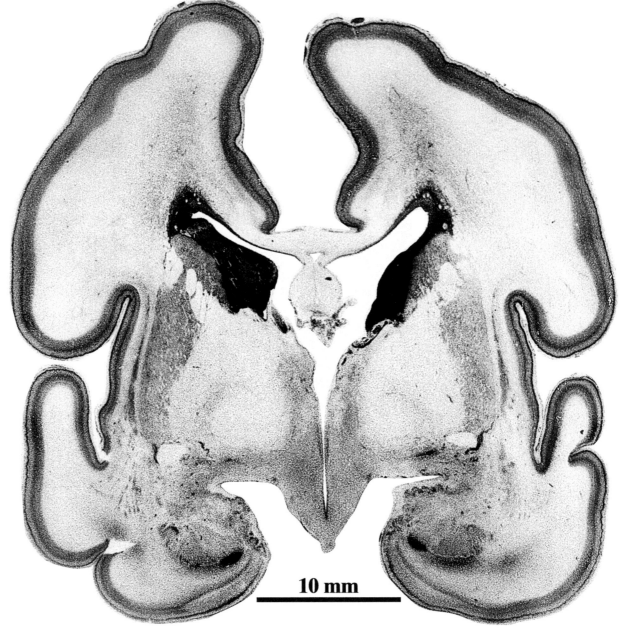

10 mm

See detail of the brain core
in Plates 43 and B.

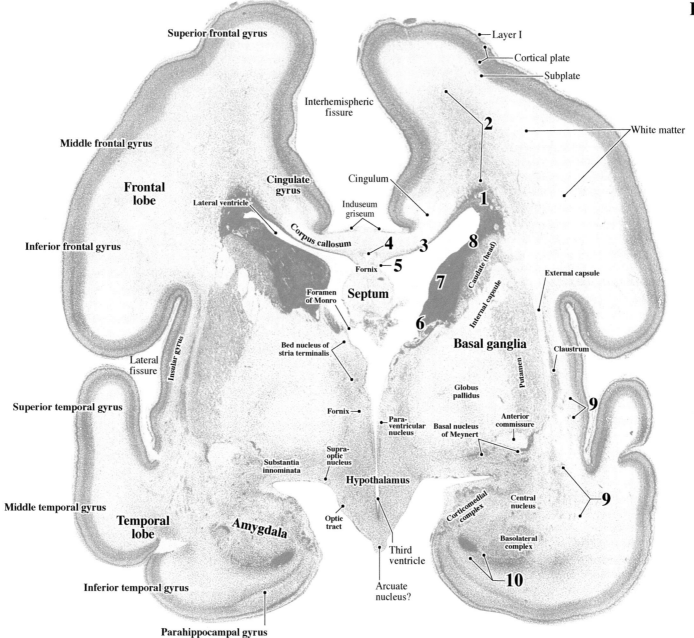

Superior frontal gyrus

Layer I

Cortical plate

Subplate

Interhemispheric
fissure

White matter

Middle frontal gyrus

Cingulum

Frontal
lobe

Cingulate
gyrus

2

Induseum
griseum

1

Lateral ventricle

Inferior frontal gyrus

Corpus callosum

4   **3**   **8**

Fornix

5

Caudate (head)

7

External capsule

Foramen
of Monro

Septum

Internal capsule

6

Basal ganglia

Bed nucleus of
stria terminalis

Claustrum

Insular gyrus

Putamen

Lateral
fissure

Globus
pallidus

Superior temporal gyrus

Fornix

Para-
ventricular
nucleus

9

Basal nucleus
of Meynert

Anterior
commissure

Substantia
innominata

Supra-
optic
nucleus

Middle temporal gyrus

Central
nucleus

9

Hypothalamus

Temporal
lobe

Amygdala

Optic
tract

Corticomedial
complex

Basolateral
complex

Inferior temporal gyrus

Third
ventricle

10

Arcuate
nucleus?

Parahippocampal gyrus

PLATE 23A
CR 260 mm
GW 30, Y14-59
Frontal
Section 821

Remnants of the germinal matrix, migratory streams, and transitional fields

1 *Frontal NEP and SVZ*

2 *Frontal STF*

3 *Callosal GEP*

4 *Callosal sling*

5 *Fornical GEP*

6 *Strionuclear GEP*

7 *Anteromedial NEP and SVZ*

8 *Parahippocampal NEP, SVZ, and STF*

9 *Temporal NEP and SVZ (intermingled with the amygdaloid G/EP)*

10 *Amygdaloid G/EP*

11 *Lateral migratory stream (cortical)*

12 *Subpial granular layer (cortical)*

GEP - Glioepithelium
G/EP - Glioepithelium/ependyma
NEP - Neuroepithelium
STF - Stratified transitional field
SVZ - Subventricular zone

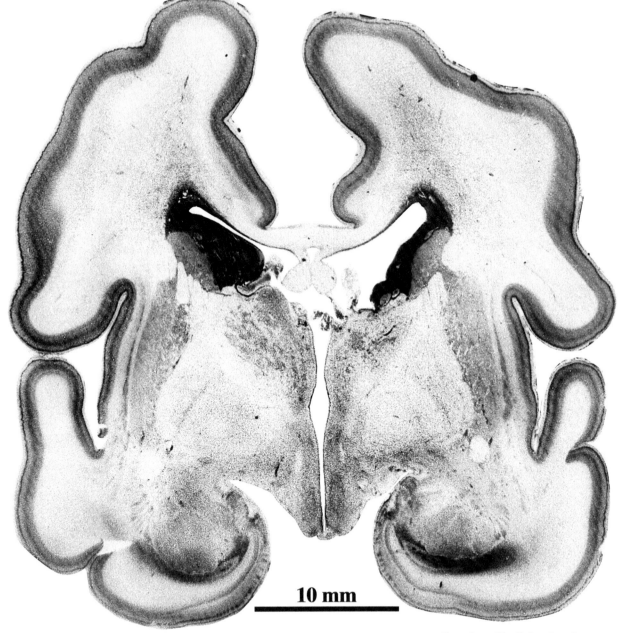

10 mm

See detail of the brain core in Plates 44 and B.

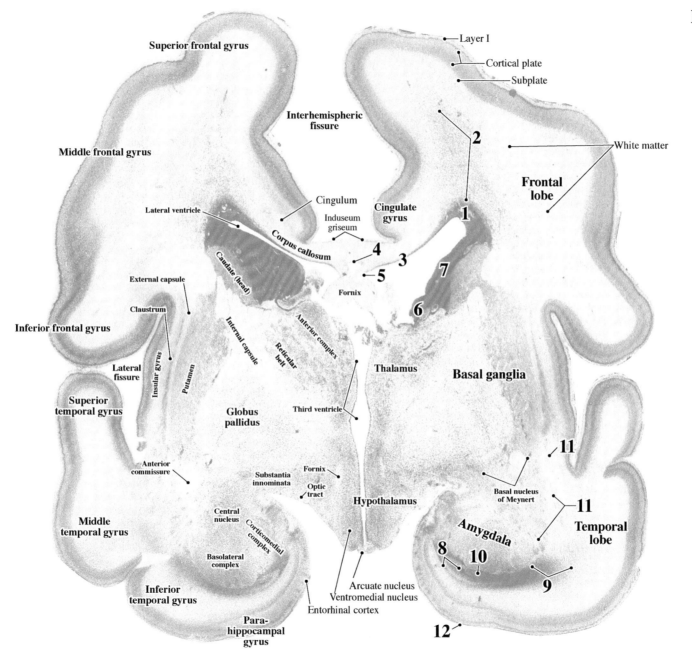

Superior frontal gyrus

Layer I

Cortical plate

Subplate

Interhemispheric fissure

Middle frontal gyrus

White matter

2

Frontal lobe

Cingulum

1

Cingulate gyrus

Lateral ventricle

Induseum griseum

Corpus callosum

4

3

External capsule

Caudate (head)

5

7

Inferior frontal gyrus

Claustrum

Internal capsule

Fornix

6

Anterior complex

Reticular belt

Lateral fissure

Insular gyrus

Putamen

Thalamus

Basal ganglia

Superior temporal gyrus

Globus pallidus

Third ventricle

11

Anterior commissure

Substantia innominata

Fornix

Basal nucleus of Meynert

Optic tract

11

Middle temporal gyrus

Central nucleus

Corticomedial complex

Hypothalamus

Amygdala

Temporal lobe

Basolateral complex

8

10

9

Inferior temporal gyrus

Arcuate nucleus

Ventromedial nucleus

Para-hippocampal gyrus

Entorhinal cortex

12

52

PLATE 24A
CR 260 mm
GW 30, Y14-59
Frontal
Section 861

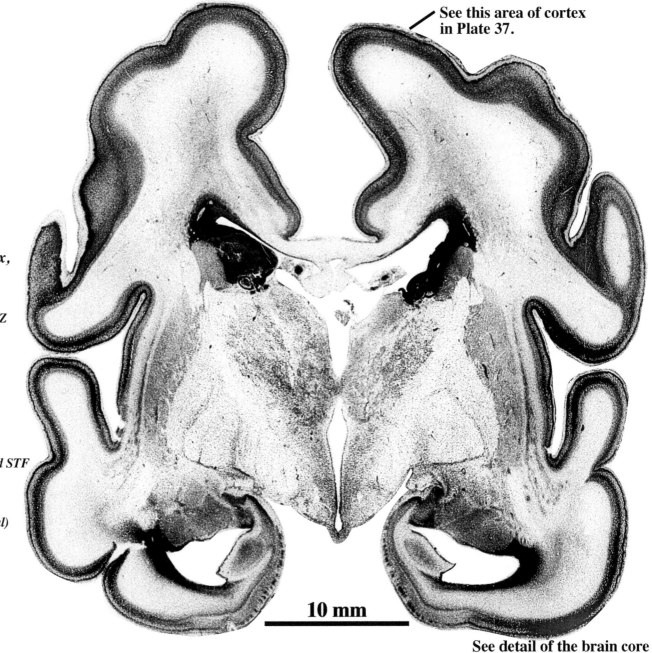

See this area of cortex
in Plate 37.

Remnants of the germinal matrix,
migratory streams, and
transitional fields

1   *Frontal/paracentral NEP and SVZ*
2   *Frontal/paracentral STF*
3   *Callosal GEP*
4   *Strionuclear GEP*
5   *Anteromedial/posterior striatal*
    NEP and SVZ
6   *Amygdaloid G/EP*
7   *Alvear glioepithelium*
8   *Parahippocampal NEP, SVZ, and STF*
9   *Temporal NEP and SVZ*
10 *Temporal STF*
11 *Lateral migratory stream (cortical)*
12 *Subpial granular layer (cortical)*

GEP - Glioepithelium
G/EP - Glioepithelium/ependyma
NEP - Neuroepithelium
STF - Stratified transitional field
SVZ - Subventricular zone

10 mm

See detail of the brain core
in Plates 45 and B.

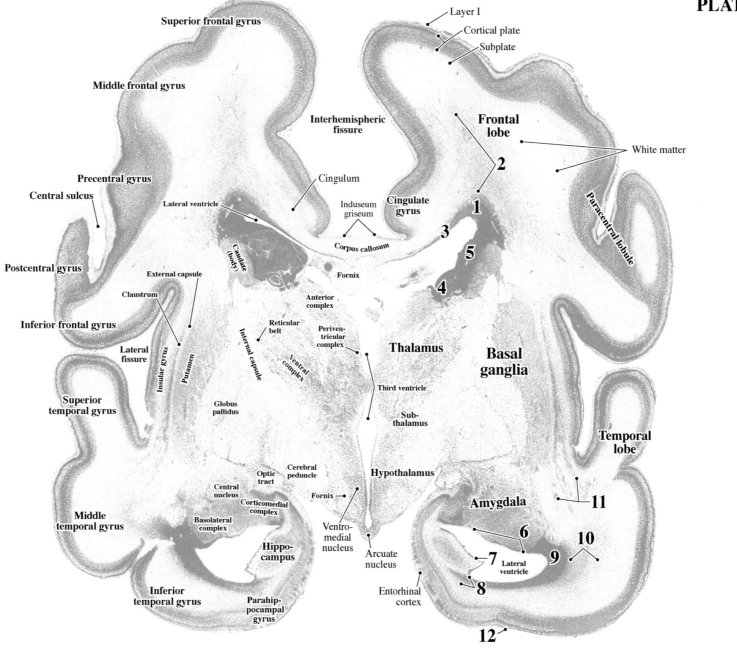

Superior frontal gyrus

Middle frontal gyrus

Interhemispheric fissure

Layer I

Cortical plate

Subplate

Frontal lobe

White matter

Precentral gyrus

Central sulcus

Cingulum

Induseum griseum

Cingulate gyrus

Paracentral lobule

Lateral ventricle

2

1

3

5

4

Postcentral gyrus

Caudate (body)

Corpus callosum

Fornix

External capsule

Claustrum

Anterior complex

Inferior frontal gyrus

Reticular belt

Periventricular complex

Thalamus

Basal ganglia

Lateral fissure

Insular gyrus

Putamen

Internal capsule

Ventral complex

Third ventricle

Superior temporal gyrus

Globus pallidus

Sub-thalamus

Temporal lobe

Optic tract

Cerebral peduncle

Hypothalamus

11

Central nucleus

Corticomedial complex

Fornix

Amygdala

Middle temporal gyrus

Basolateral complex

6

10

Hippo-campus

Ventro-medial nucleus

7

Lateral ventricle

9

8

Inferior temporal gyrus

Parahip-pocampal gyrus

Arcuate nucleus

Entorhinal cortex

12

PLATE 25A
CR 260 mm
GW 30, Y14-59
Frontal
Section 901

Remnants of the germinal matrix, migratory streams, and transitional fields

1 *Paracentral NEP and SVZ*

2 *Paracentral STF*

3 *Callosal GEP*

4 *Posterior striatal NEP and SVZ*

5 *Strionuclear GEP*

6 *Amygdaloid GE/P*

7 *Alvear GEP*

8 *Parahippocampal NEP, SVZ, and STF*

9 *Temporal NEP and SVZ*

10 *Temporal SEF*

11 *Lateral migratory stream (cortical)*

12 *Subpial granular layer (cortical)*

GEP - Glioepithelium
G/EP - Glioepithelium/ependyma
NEP - Neuroepithelium
STF - Stratified transitional field
SVZ - Subventricular zone

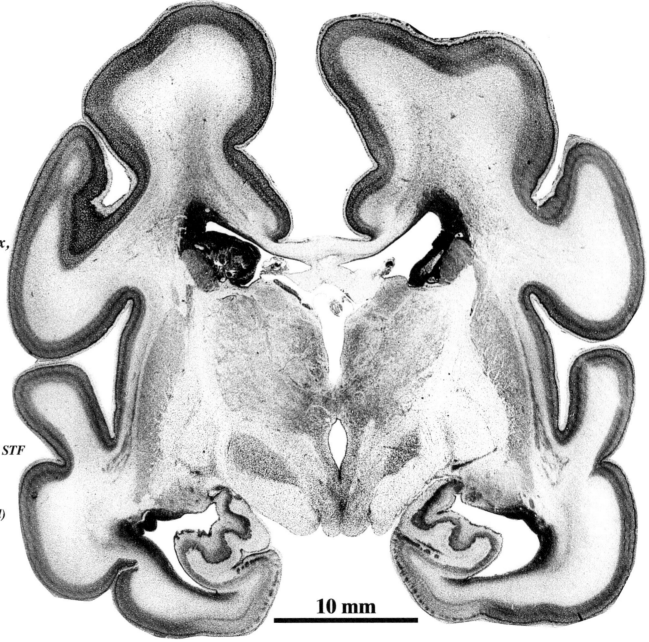

10 mm

See detail of the brain core in Plates 46 and B.

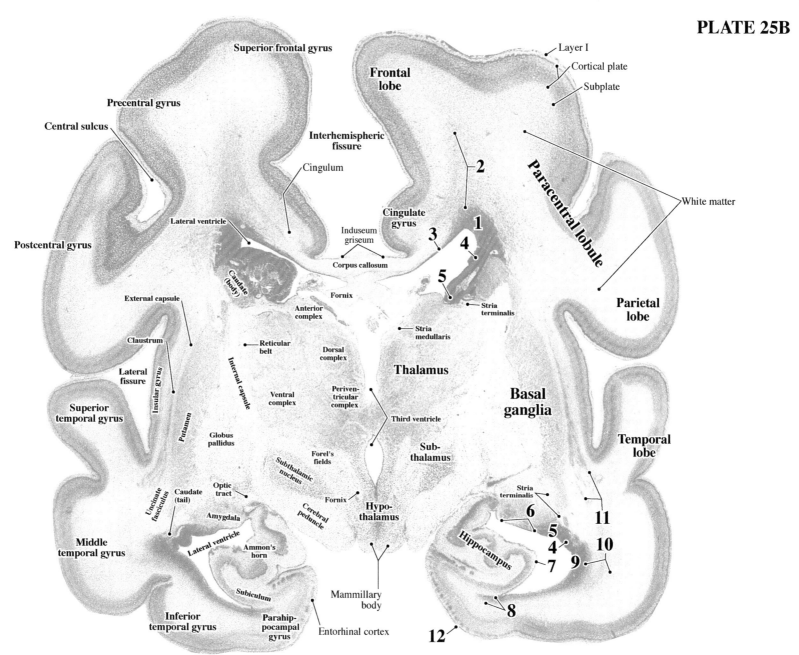

Superior frontal gyrus

Frontal lobe

Layer I

Cortical plate

Subplate

Precentral gyrus

Interhemispheric fissure

Central sulcus

Cingulum

2

Paracentral lobule

Cingulate gyrus

White matter

Lateral ventricle

1

Induseum griseum

3

4

Postcentral gyrus

5

Corpus callosum

Parietal lobe

Fornix

Stria terminalis

Caudate (body)

Anterior complex

Stria medullaris

External capsule

Reticular belt

Dorsal complex

Claustrum

Thalamus

Lateral fissure

Internal capsule

Ventral complex

Periventricular complex

Basal ganglia

Superior temporal gyrus

Insular gyrus

Third ventricle

Temporal lobe

Putamen

Globus pallidus

Forel's fields

Sub- thalamus

Subthalamic nucleus

Stria terminalis

Uncinate fasciculus

Caudate (tail)

Optic tract

Fornix

6

11

Amygdala

Cerebral peduncle

Hypo- thalamus

5

Middle temporal gyrus

Lateral ventricle

Ammon's horn

Hippocampus

4

10

9

7

Subiculum

8

Mammillary body

Inferior temporal gyrus

Parahip- pocampal gyrus

Entorhinal cortex

12

PLATE 26A
CR 260 mm
GW 30, Y14-59
Frontal
Section 981

Remnants of the germinal matrix, migratory streams, and transitional fields

1 *Paracentral NEP and SVZ*

2 *Paracentral STF*

3 *Callosal GEP*

4 *Posterior striatal NEP and SVZ*

5 *Strionuclear GEP*

6 *Alvear GEP*

7 *Subgranular zone (dentate)*

8 *Parahippocampal NEPm, SVZ, and STF*

9 *Temporal NEP and SVZ*

10 *Temporal STF*

11 *Lateral migratory stream (cortical)*

12 *Subpial granular layer (cortical)*

GEP - Glioepithelium
G/EP - Glioepithelium/ependyma
NEP - Neuroepithelium
STF - Stratified transitional field
SVZ - Subventricular zone

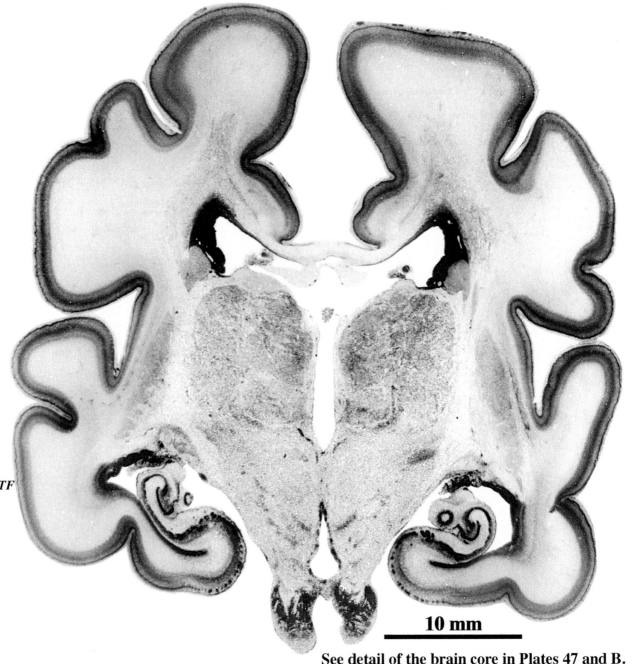

10 mm

See detail of the brain core in Plates 47 and B.

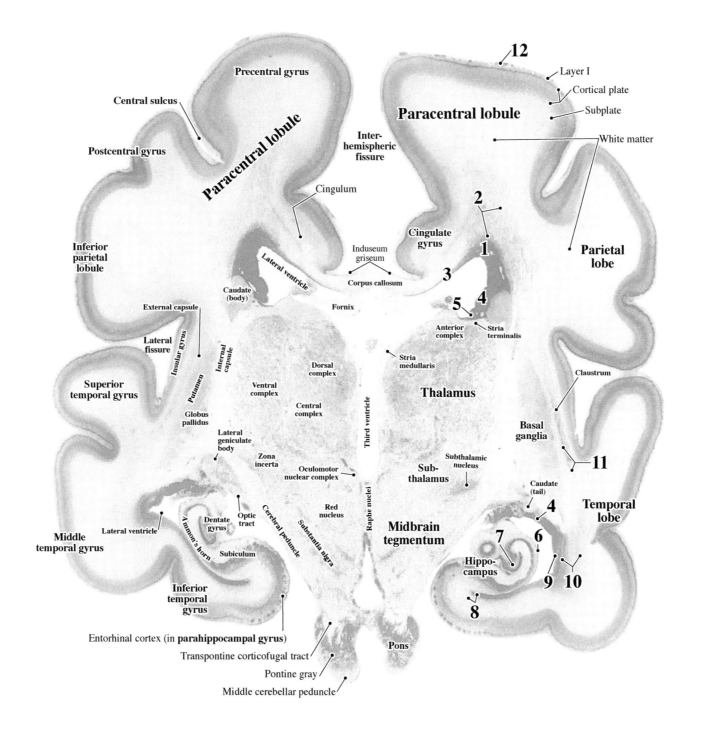

12

Layer I

Cortical plate

Subplate

White matter

Precentral gyrus

Central sulcus

Paracentral lobule

Postcentral gyrus

Paracentral lobule

Inter-
hemispheric
fissure

Cingulum

2

**Cingulate
gyrus**

**Parietal
lobe**

Inferior
parietal
lobule

1

Induseum
griseum

Lateral ventricle

3

Caudate
(body)

Corpus callosum

4

5

External capsule

Fornix

Anterior
complex

Stria
terminalis

Lateral
fissure

Internal
capsule

Insular gyrus

Stria
medullaris

Claustrum

Dorsal
complex

**Superior
temporal gyrus**

Putamen

Ventral
complex

Thalamus

Globus
pallidus

Central
complex

Third ventricle

Basal
ganglia

Lateral
geniculate
body

11

Zona
incerta

Oculomotor
nuclear complex

Subthalamic
nucleus

Sub-
thalamus

Caudate
(tail)

4

**Temporal
lobe**

Red
nucleus

Raphe nuclei

Optic
tract

Dentate
gyrus

Cerebral peduncle

7

6

Ammon's horn

Lateral ventricle

Substantia nigra

**Midbrain
tegmentum**

**Middle
temporal gyrus**

Subiculum

Hippo-
campus

9

10

**Inferior
temporal
gyrus**

8

Entorhinal cortex (in **parahippocampal gyrus**)

Transpontine corticofugal tract

Pontine gray

Middle cerebellar peduncle

Pons

58

PLATE 27A
CR 260 mm
GW 30, Y14-59
Frontal
Section 1021

Remnants of the germinal matrix,
migratory streams, and
transitional fields

1   *Paracentral/parietal NEP and SVZ*
2   *Paracentral/parietal STF*
3   *Callosal GEP*
4   *Posterior striatal NEP and SVZ*
5   *Strionuclear GEP*
6   *Alvear GEP*
7   *Subgranular zone (dentate)*
8   *Parahippocampal NEP, SVZ, and STF*
9   *Temporal NEP and SVZ*
10   *Temporal STF*
11   *Lateral migratory stream (cortical)*
12   *Subpial granular layer (cortical)*

GEP - Glioepithelium
NEP - Neuroepithelium
STF - Stratified transitional field
SVZ - Subventricular zone

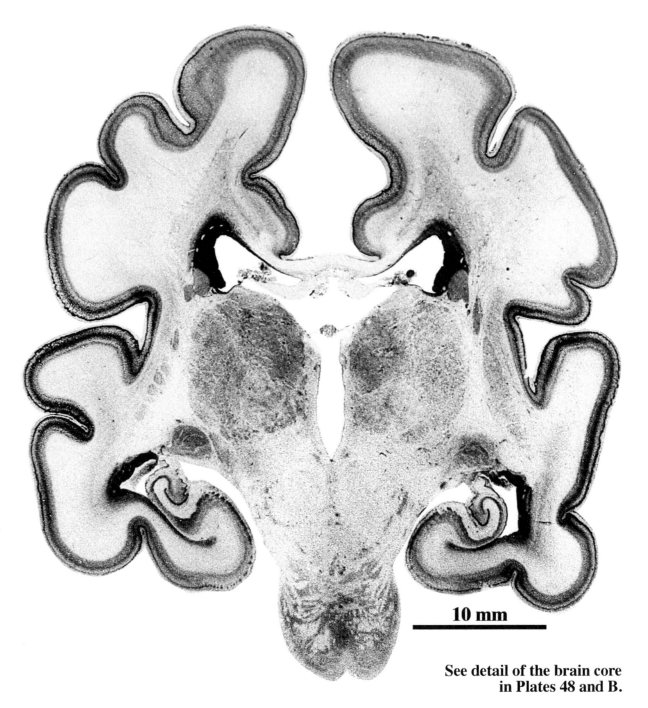

10 mm

See detail of the brain core
in Plates 48 and B.

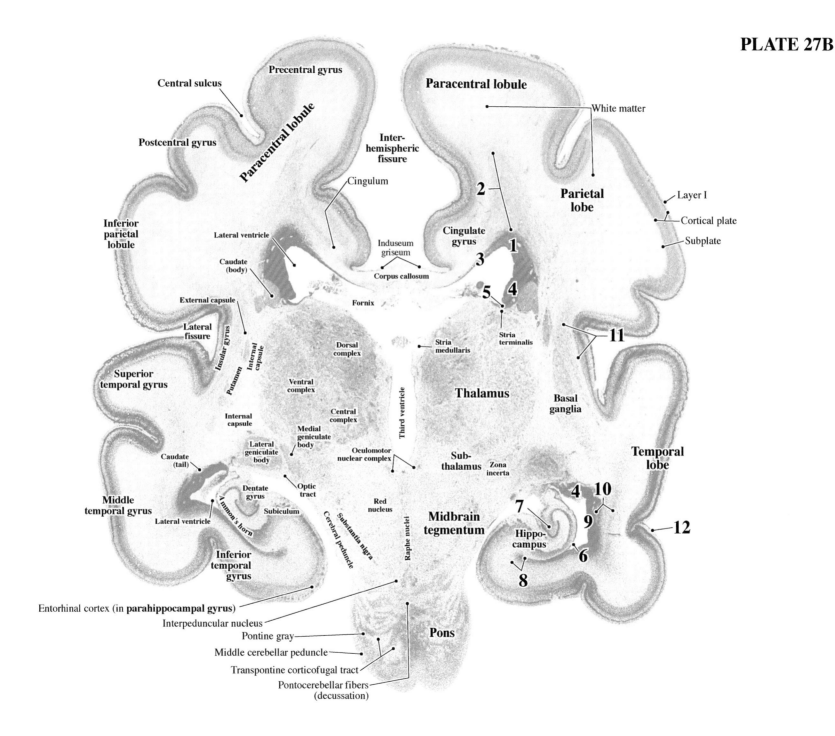

Central sulcus

Precentral gyrus

Paracentral lobule

Paracentral lobule

White matter

Postcentral gyrus

Inter-hemispheric fissure

Cingulum

Parietal lobe

Layer I

Inferior parietal lobule

Lateral ventricle

Induseum griseum

Cingulate gyrus

Cortical plate

Subplate

Caudate (body)

Corpus callosum

1

2

3

External capsule

Fornix

5 **4**

11

Lateral fissure

Internal capsule

Dorsal complex

Stria medullaris

Stria terminalis

Superior temporal gyrus

Insular gyrus

Putamen

Ventral complex

Thalamus

Basal ganglia

Internal capsule

Central complex

Third ventricle

Medial geniculate body

Oculomotor nuclear complex

Sub-thalamus

Zona incerta

Temporal lobe

Caudate (tail)

Lateral geniculate body

Optic tract

Red nucleus

4 **10**

Middle temporal gyrus

Dentate gyrus

Ammon's horn

Subiculum

Substantia nigra

Cerebral peduncle

Raphe nuclei

Midbrain tegmentum

7

9

Lateral ventricle

Hippo-campus

6

Inferior temporal gyrus

8

12

Entorhinal cortex (in **parahippocampal gyrus**)

Interpeduncular nucleus

Pontine gray

Middle cerebellar peduncle

Transpontine corticofugal tract

Pontocerebellar fibers (decussation)

Pons

PLATE 28A
CR 260 mm
GW 30, Y14-59
Frontal
Section 1061

Remnants of the germinal matrix,
migratory streams, and
transitional fields

1 *Paracentral/parietal NEP and SVZ*
2 *Paracentral/parietal STF*
3 *Callosal GEP*
4 *Posterior striatal NEP and SVZ*
5 *Strionuclear GEP*
6 *Alvear GEP*
7 *Subgranular zone (dentate)*
8 *Parahippocampal NEP, SVZ, and STF*
9 *Temporal NEP and SVZ*
10 *Temporal STF*
11 *Mesencephalic G/EP*
12 *Subpial granular layer (cortical)*

GEP - Glioepithelium
G/EP - Glioepithelium/ependyma
NEP - Neuroepithelium
STF - Stratified transitional field
SVZ - Subventricular zone

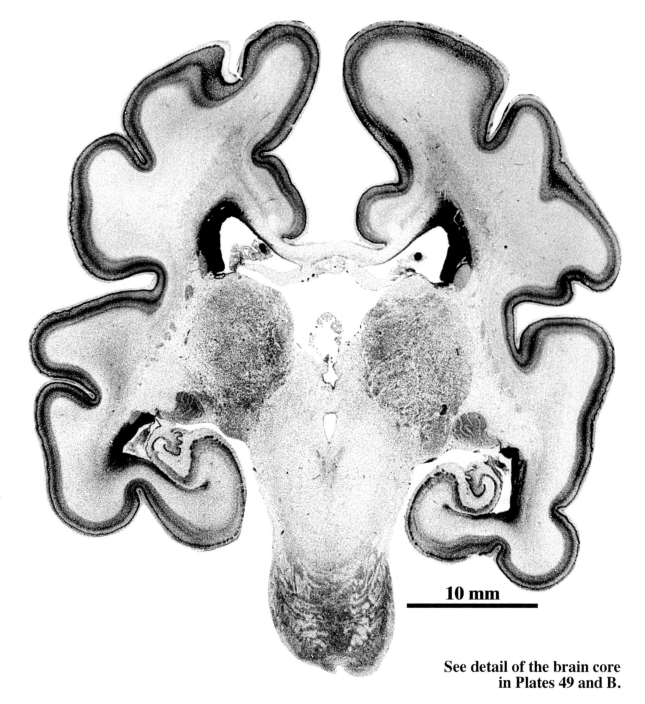

10 mm

See detail of the brain core
in Plates 49 and B.

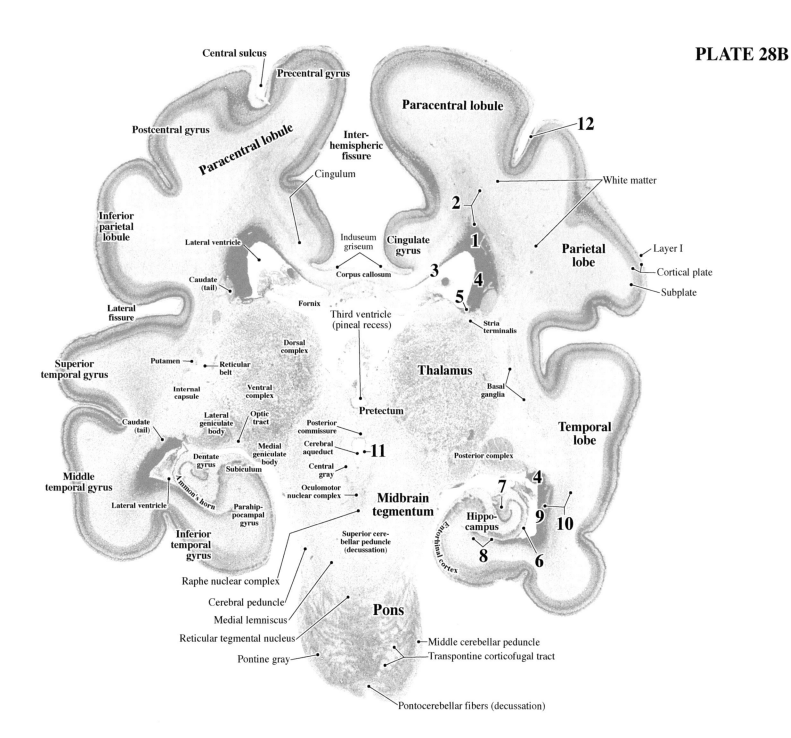

Central sulcus

Precentral gyrus

Paracentral lobule

12

Postcentral gyrus

Inter-
hemispheric
fissure

Paracentral lobule

Cingulum

White matter

Inferior
parietal
lobule

Induseum
griseum

Cingulate
gyrus

2

1

Parietal
lobe

Layer I

Lateral ventricle

3

Cortical plate

Corpus callosum

4

Subplate

Caudate
(tail)

5

Lateral
fissure

Fornix

Third ventricle
(pineal recess)

Stria
terminalis

Dorsal
complex

Thalamus

Superior
temporal gyrus

Putamen

Reticular
belt

Basal
ganglia

Internal
capsule

Ventral
complex

Optic
tract

Pretectum

Caudate
(tail)

Lateral
geniculate
body

Posterior
commissure

Posterior complex

Temporal
lobe

Dentate
gyrus

Medial
geniculate
body

Cerebral
aqueduct

11

Middle
temporal
gyrus

Subiculum

Central
gray

4

Ammon's horn

Parahip-
pocampal
gyrus

Oculomotor
nuclear complex

7

Lateral ventricle

Midbrain
tegmentum

Hippo-
campus

9

10

Inferior
temporal
gyrus

Superior cere-
bellar peduncle
(decussation)

Entorhinal cortex

8

6

Raphe nuclear complex

Cerebral peduncle

Medial lemniscus

Pons

Reticular tegmental nucleus

Middle cerebellar peduncle

Pontine gray

Transpontine corticofugal tract

Pontocerebellar fibers (decussation)

PLATE 29A
CR 260 mm
GW 30, Y14-59
Frontal
Section 1101

Remnants of the germinal matrix,
migratory streams, and
transitional fields

1   *Paracentral/parietal NEP and SVZ*
2   *Paracentral/parietal STF*
3   *Callosal GEP*
4   *Posterior striatal NEP and SVZ*
5   *Strionuclear GEP*
6   *Alvear GEP*
7   *Subgranular zone (dentate)*
8   *Parahippocampal NEP, SVZ, and STF*
9   *Temporal NEP and SVZ*
10   *Temporal STF*
11   *Mesencephalic G/EP*
12   *Subpial granular layer (cortical)*

GEP - Glioepithelium
G/EP - Glioepithelium/ependyma
NEP - Neuroepithelium
STF - Stratified transitional field
SVZ - Subventricular zone

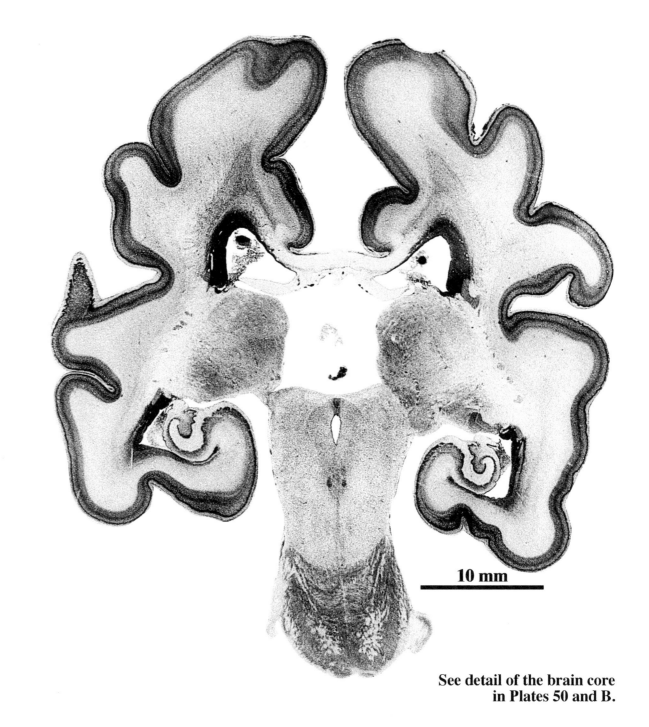

10 mm

See detail of the brain core
in Plates 50 and B.

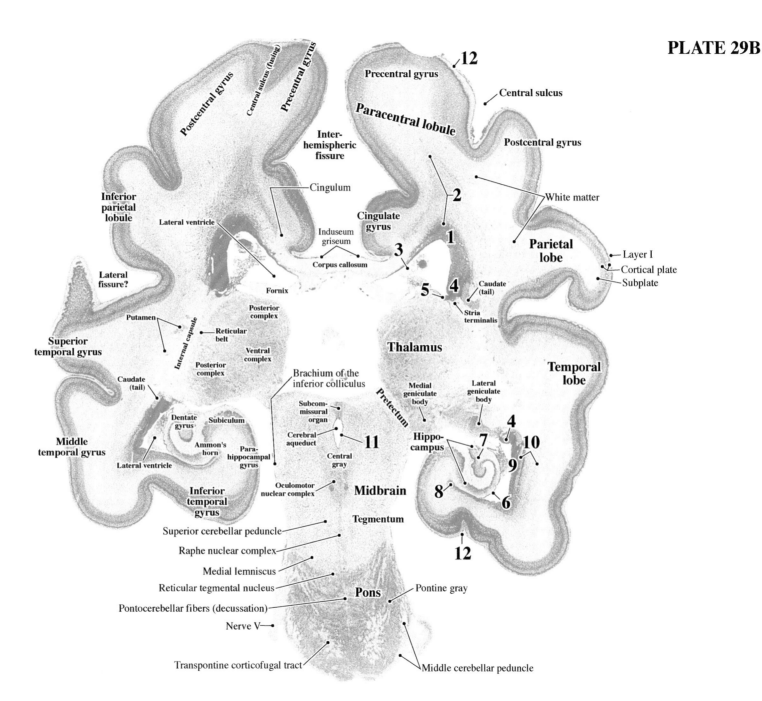

Postcentral gyrus

Central sulcus (fusing)

Precentral gyrus

Precentral gyrus

12

Central sulcus

Paracentral lobule

Postcentral gyrus

Inter-hemispheric fissure

Cingulum

White matter

2

Inferior parietal lobule

Lateral ventricle

Cingulate gyrus

Induseum griseum

1

Parietal lobe

3

Corpus callosum

4

Layer I

Cortical plate

Subplate

Lateral fissure?

Fornix

5

Caudate (tail)

Stria terminalis

Posterior complex

Putamen

Reticular belt

Internal capsule

Ventral complex

Thalamus

Temporal lobe

Superior temporal gyrus

Caudate (tail)

Posterior complex

Brachium of the inferior colliculus

Medial geniculate body

Lateral geniculate body

4

Dentate gyrus

Subiculum

Subcommissural organ

Pretectum

Hippo-campus

7

10

Middle temporal gyrus

Ammon's horn

Cerebral aqueduct

11

9

Lateral ventricle

Para-hippocampal gyrus

Central gray

8

6

Inferior temporal gyrus

Oculomotor nuclear complex

Midbrain

12

Superior cerebellar peduncle

Tegmentum

Raphe nuclear complex

Medial lemniscus

Reticular tegmental nucleus

Pontine gray

Pontocerebellar fibers (decussation)

Pons

Nerve V

Transpontine corticofugal tract

Middle cerebellar peduncle

PLATE 30A
CR 260 mm
GW 30, Y14-59
Frontal
Section 1161

Remnants of the germinal matrix,
migratory streams, and
transitional fields

1 *Paracentral/parietal NEP and SVZ*
2 *Paracentral/parietal STF*
3 *Callosal GEP*
4 *Posterior striatal NEP and SVZ*
5 *Strionuclear GEP*
6 *Alvear GEP*
7 *Subgranular zone (dentate)*
8 *Parahippocampal NEP, SVZ, and SYF*
9 *Temporal NEP and SVZ*
10 *Temporal STF*
11 *Mesencephalic G/EP*
12 *Raphe migration*
13 *External germinal layer (cerebellum)*
14 *Subpial granular layer (cortical)*

GEP - Glioepithelium
G/EP - Glioepithelium/ependyma
NEP - Neuroepithelium
STF - Stratified transitional field
SVZ - Subventricular zone

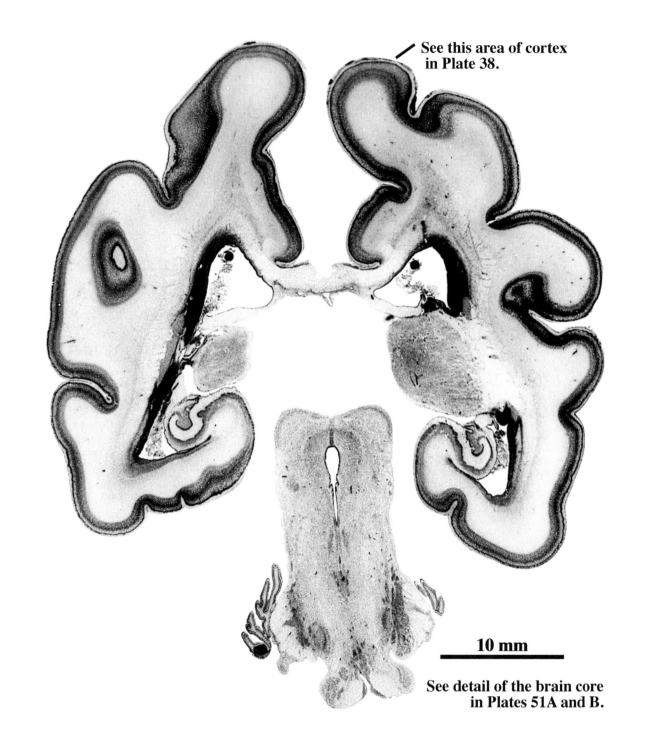

See this area of cortex
in Plate 38.

10 mm

See detail of the brain core
in Plates 51A and B.

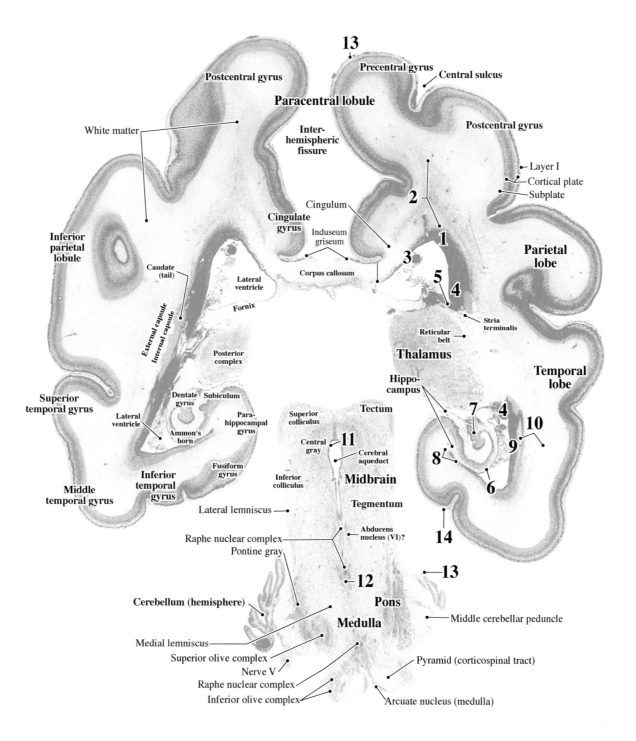

13

Postcentral gyrus

Paracentral lobule

Precentral gyrus — Central sulcus

White matter

Inter-
hemispheric
fissure

Postcentral gyrus

Layer I
Cortical plate
Subplate

Cingulum

2

Cingulate
gyrus

Induseum
griseum

1

**Parietal
lobe**

Inferior
parietal
lobule

Caudate
(tail)

Lateral
ventricle

Corpus callosum

3

5

4

Fornix

Stria
terminalis

Reticular
belt

External capsule
Internal capsule

Posterior
complex

Thalamus

**Temporal
lobe**

Hippo-
campus

**Superior
temporal gyrus**

Dentate
gyrus

Subiculum

Lateral
ventricle

Para-
hippocampal
gyrus

7

4

10

Ammon's
horn

Tectum

Superior
colliculus

9

8

Central
gray

11

Cerebral
aqueduct

Fusiform
gyrus

6

**Inferior
temporal
gyrus**

Inferior
colliculus

Midbrain

**Middle
temporal gyrus**

Tegmentum

Lateral lemniscus

14

Abducens
nucleus (VI)?

Raphe nuclear complex

Pontine gray

13

12

Pons

Cerebellum (hemisphere)

Medulla

Middle cerebellar peduncle

Medial lemniscus

Superior olive complex

Nerve V

Pyramid (corticospinal tract)

Raphe nuclear complex

Inferior olive complex

Arcuate nucleus (medulla)

PLATE 31A
CR 260 mm
GW 30, Y14-59
Frontal
Section 1201

Remnants of the germinal matrix,
migratory streams, and
transitional fields

1 *Paracentral/parietal NEP and SVZ*
2 *Paracentral/parietal STF*
3 *Callosal GEP*
4 *Posterior striatal NEP and SVZ*
5 *Strionuclear GEP*
6 *Alvear GEP*
7 *Subgranular zone (dentate)*
8 *Parahippocampal NEP, SVZ, and STF*
9 *Temporal NEP and SVZ*
10 *Temporal STF*
11 *Mesencephalic G.EP*
12 *Isthmal and pontine G/EP*
13 *Raphe migration*
14 *External germinal layer (cerebellum)*
15 *Subpial granular layer (cortical)*

GEP - Glioepithelium
G/EP - Glioepithelium/ependyma
NEP - Neuroepithelium
STF - Stratified transitional field
SVZ - Subventricular zone

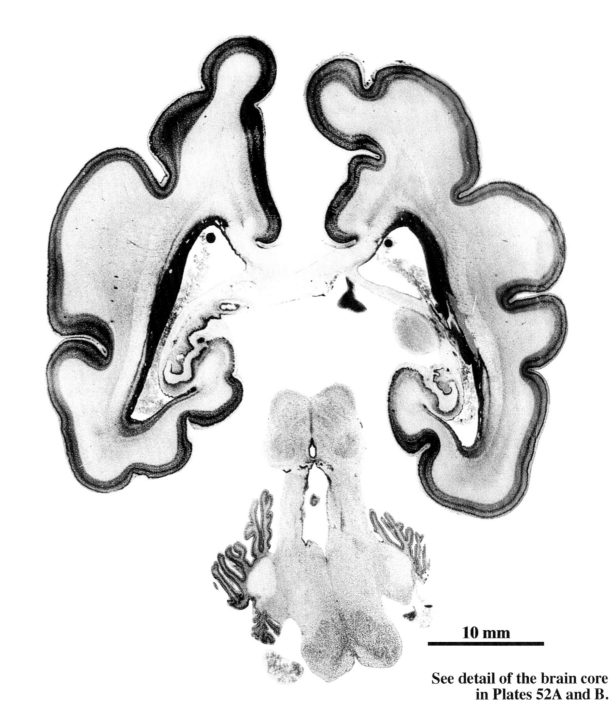

10 mm

See detail of the brain core
in Plates 52A and B.

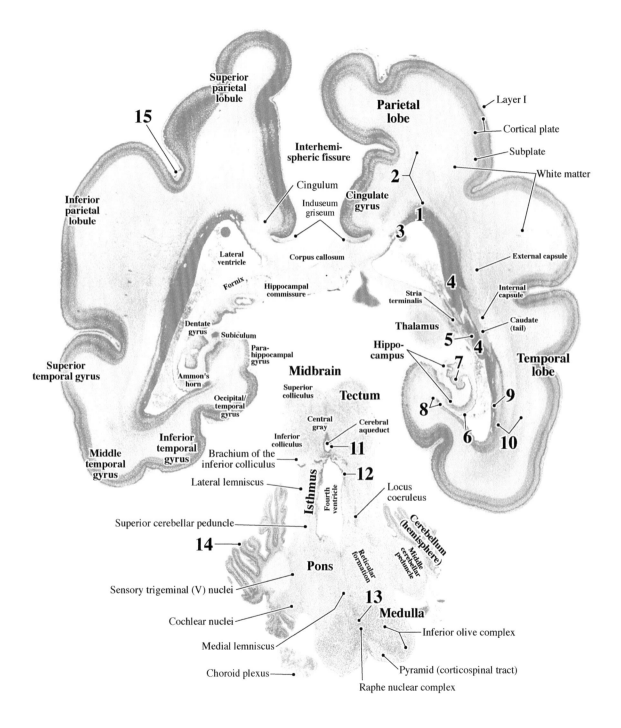

Superior parietal lobule

15

Parietal lobe

Layer I

Cortical plate

Subplate

White matter

Interhemispheric fissure

Cingulum

2

Induseum griseum

Cingulate gyrus

Inferior parietal lobule

1

3

External capsule

Lateral ventricle

Corpus callosum

Internal capsule

Fornix

Hippocampal commissure

4

Stria terminalis

Caudate (tail)

Dentate gyrus

Subiculum

Thalamus

Para-hippocampal gyrus

Hippocampus

5

4

Superior temporal gyrus

Ammon's horn

Occipital/temporal gyrus

Midbrain

Superior colliculus

7

9

Temporal lobe

8

Tectum

6

10

Inferior colliculus

Central gray

Cerebral aqueduct

Inferior temporal gyrus

Middle temporal gyrus

Brachium of the inferior colliculus

11

Lateral lemniscus

12

Isthmus

Fourth ventricle

Locus coeruleus

Superior cerebellar peduncle

Cerebellum (hemisphere)

Middle cerebellar peduncle

14

Pons

Reticular formation

Sensory trigeminal (V) nuclei

13

Medulla

Cochlear nuclei

Inferior olive complex

Medial lemniscus

Pyramid (corticospinal tract)

Choroid plexus

Raphe nuclear complex

PLATE 32A
CR 260 mm
GW 30, Y14-59
Frontal
Section 1261

Remnants of the germinal matrix, migratory streams, and transitional fields

1 *Parietal NEP and SVZ*

2 *Parietal STF*

3 *Callosal GEP*

4 *Fimbrial GEP*

5 *Posterior striatal NEP and SVZ*

6 *Alvear GEP*

7 *Subgranular zone (dentate)*

8 *Occipital NEP and SVZ?*

9 *Occipital STF?*

10 *Temporal NEP and STF*

11 *Temporal STF*

12 *Pontine and medullary G/EP*

13 *External germinal layer (cerebellum)*

14 *Subpial granular layer (cortical)*

GEP - Glioepithelium
G/EP - Glioepithelium/ependyma
NEP - Neuroepithelium
STF - Stratified transitional field
SVZ - Subventricular zone

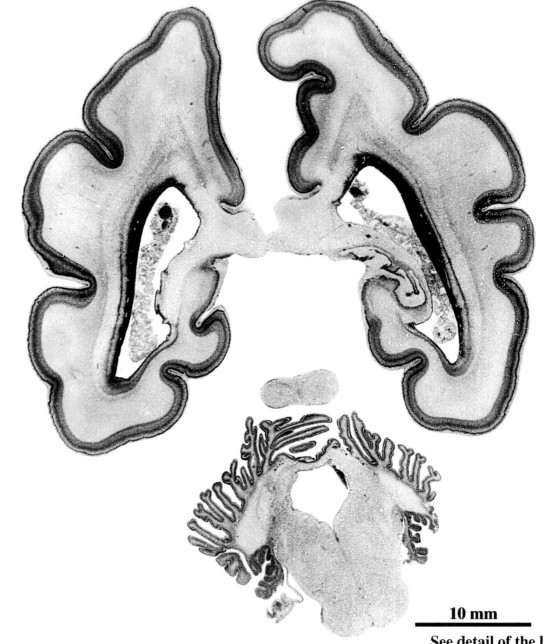

10 mm

See detail of the brain core in Plates 53A and B.

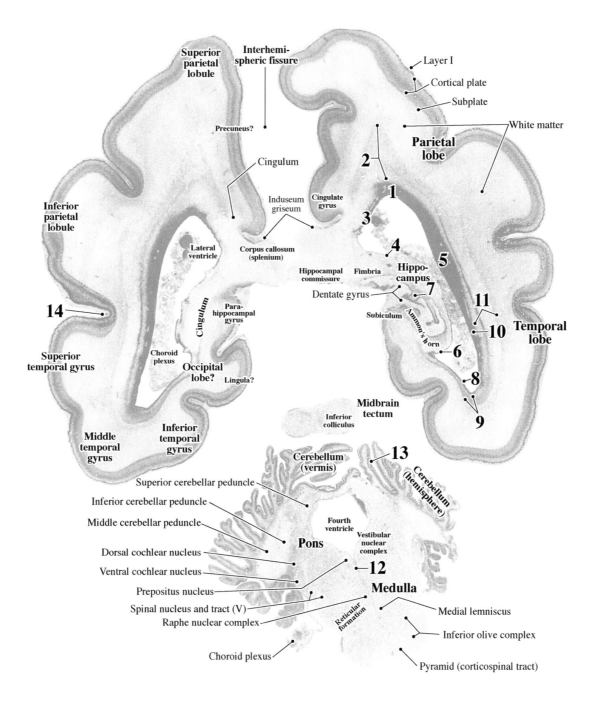

Superior parietal lobule

Interhemi- spheric fissure

Layer I

Cortical plate

Subplate

White matter

Precuneus?

Parietal lobe

Cingulum

2

Induseum griseum

Cingulate gyrus

1

Inferior parietal lobule

3

4

5

Corpus callosum (splenium)

Lateral ventricle

Hippocampal commissure

Fimbria

Hippo- campus

7

11

Dentate gyrus

14

Cingulum

Para- hippocampal gyrus

Subiculum

Ammon's horn

10

Temporal lobe

6

Superior temporal gyrus

Choroid plexus

Occipital lobe?

Lingula?

8

9

Midbrain tectum

Inferior colliculus

Middle temporal gyrus

Inferior temporal gyrus

Cerebellum (vermis)

13

Cerebellum (hemisphere)

Superior cerebellar peduncle

Inferior cerebellar peduncle

Middle cerebellar peduncle

Dorsal cochlear nucleus

Ventral cochlear nucleus

Prepositus nucleus

Spinal nucleus and tract (V)

Raphe nuclear complex

Choroid plexus

Fourth ventricle

Vestibular nuclear complex

Pons

12

Medulla

Reticular formation

Medial lemniscus

Inferior olive complex

Pyramid (corticospinal tract)

PLATE 33A
CR 260 mm
GW 30, Y14-59
Frontal
Section 1301

Remnants of the germinal matrix,
migratory streams, and
transitional fields

1 *Parietal NEP and SVZ*

2 *Parietal STF*

3 *Callosal GEP*

4 *Fimbrial GEP*

5 *Occipital NEP and SVZ*

6 *Occipital STF*

7 *Temporal NEP and SVZ*

8 *Temporal STF*

9 *Medullary G/EP*

10 *Germinal trigone (cerebellum)*

11 *External germinal layer (cerebellum)*

12 *Subpial granular layer (cortical)*

GEP - Glioepithelium
G/EP - Glioepithelium/ependyma
NEP - Neuroepithelium
STF - Stratified transitional field
SVZ - Subventricular zone

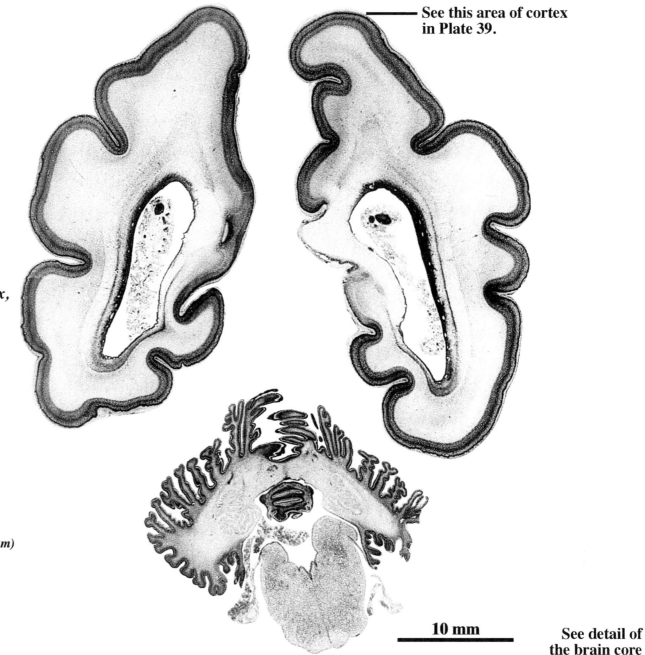

See this area of cortex
in Plate 39.

10 mm

See detail of
the brain core
in Plates 54A and B.

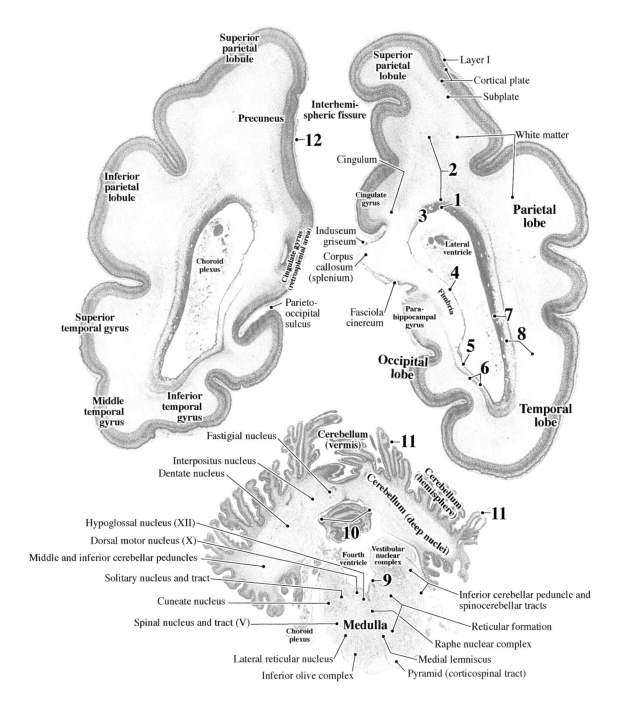

Superior
parietal
lobule

Superior
parietal
lobule

Layer I

Cortical plate

Subplate

Interhemi-
spheric fissure

Precuneus

12

White matter

Cingulum

2

Inferior
parietal
lobule

Cingulate
gyrus

1

Parietal
lobe

3

Induseum
griseum

Lateral
ventricle

Choroid
plexus

Corpus
callosum
(splenium)

Cingulate gyrus
(retrosplenial area)

4

Fimbria

Superior
temporal gyrus

Parieto-
occipital
sulcus

Fasciola
cinereum

Para-
hippocampal
gyrus

7

8

5

Occipital
lobe

6

Temporal
lobe

Middle
temporal
gyrus

Inferior
temporal
gyrus

Fastigial nucleus

Cerebellum
(vermis)

11

Interpositus nucleus

Dentate nucleus

Cerebellum
(hemisphere)

Cerebellum (deep nuclei)

11

Hypoglossal nucleus (XII)

10

Dorsal motor nucleus (X)

Middle and inferior cerebellar peduncles

Fourth
ventricle

Vestibular
nuclear
complex

Solitary nucleus and tract

9

Inferior cerebellar peduncle and
spinocerebellar tracts

Cuneate nucleus

Reticular formation

Spinal nucleus and tract (V)

Choroid
plexus

Medulla

Raphe nuclear complex

Lateral reticular nucleus

Medial lemniscus

Inferior olive complex

Pyramid (corticospinal tract)

PLATE 34A
CR 260 mm
GW 30, Y14-59
Frontal
Section 1361

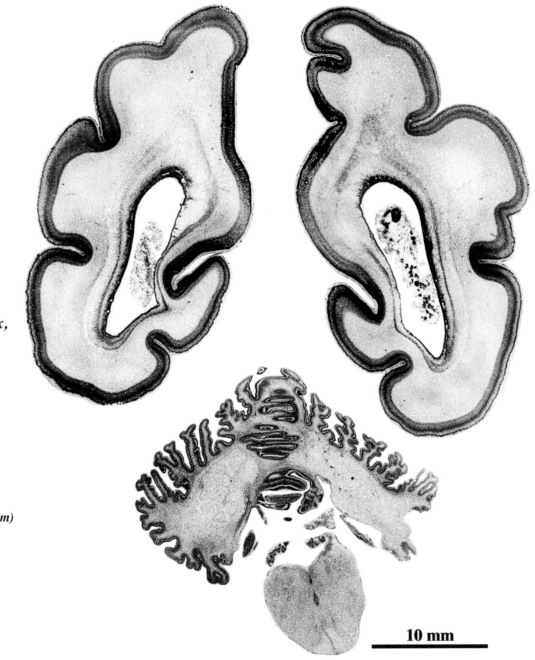

Remnants of the germinal matrix,
migratory streams, and
transitional fields

1 *Parietal NEP and SVZ*

2 *Parietal STF*

3 *Callosal GEP*

4 *Occipital NEP and SVZ*

5 *Occipital STF*

6 *Temporal NEP and SVZ*

7 *Temporal STF*

8 *External germinal layer (cerebellum)*

9 *Spinomedullary G/EP*

10 *Subpial granular layer (cortical)*

GEP - Glioepithelium
G/EP - Glioepithelium/ependyma
NEP - Neuroepithelium
STF - Stratified transitional field
SVZ - Subventricular zone

10 mm

See detail of
the brain core
in Plates 55A and B.

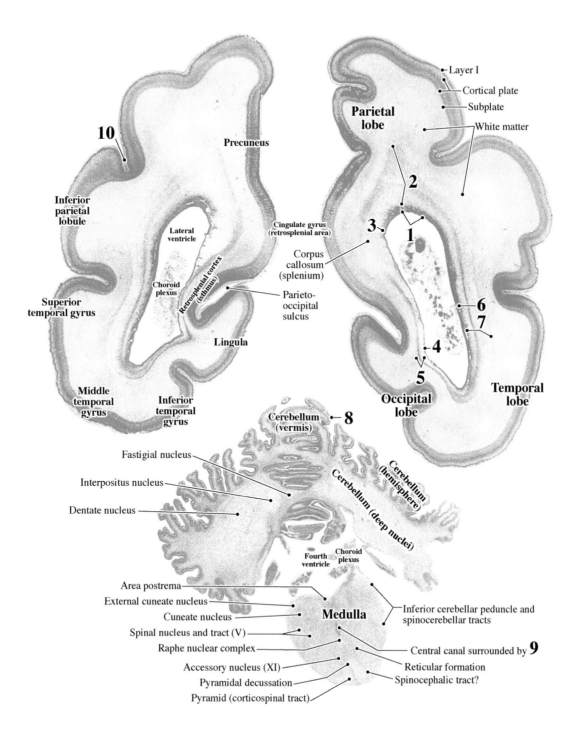

Layer I
Cortical plate
Subplate
White matter

Parietal lobe

10

Precuneus

2

Inferior parietal lobule

Lateral ventricle

Cingulate gyrus (retrosplenial area)

3

1

Corpus callosum (splenium)

Choroid plexus

Retrosplenial cortex (isthmus)

Superior temporal gyrus

Parieto-occipital sulcus

6

7

Lingula

4

5

Middle temporal gyrus

Inferior temporal gyrus

Occipital lobe

Temporal lobe

Cerebellum (vermis) **8**

Fastigial nucleus

Interpositus nucleus

Dentate nucleus

Cerebellum (hemisphere)

Cerebellum (deep nuclei)

Fourth ventricle

Choroid plexus

Area postrema

External cuneate nucleus

Cuneate nucleus

Spinal nucleus and tract (V)

Raphe nuclear complex

Accessory nucleus (XI)

Pyramidal decussation

Pyramid (corticospinal tract)

Medulla

Inferior cerebellar peduncle and spinocerebellar tracts

Central canal surrounded by **9**

Reticular formation

Spinocephalic tract?

74

PLATE 35A
CR 260 mm
GW 30, Y14-59
Frontal
Section 1521

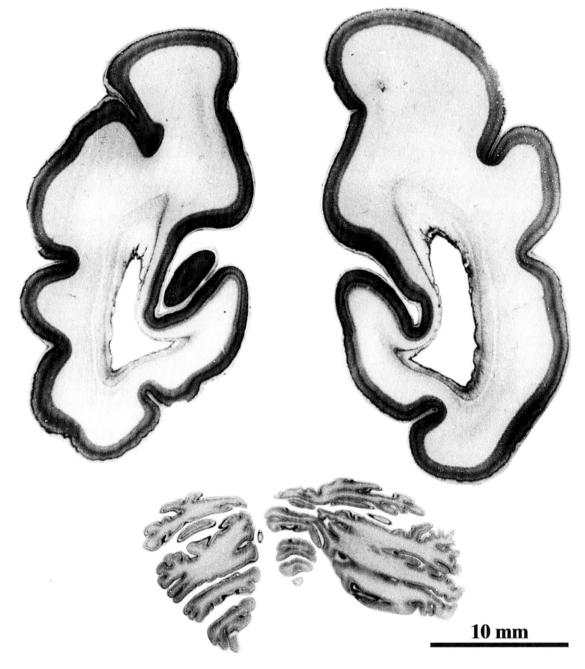

Remnants of the germinal matrix,
migratory streams, and
transitional fields

1   *Parietal NEP and SVZ*

2   *Parietal STF*

3   *Occipital NEP and SVZ*

4   *Occipital STF*

5   *Temporal NEP and SVZ?*

6   *Temporal STF?*

7   *External germinal layer (cerebellum)*

8   *Subpial granular layer (cortical)*

NEP - Neuroepithelium
STF - Stratified transitional field
SVZ - Subventricular zone

10 mm

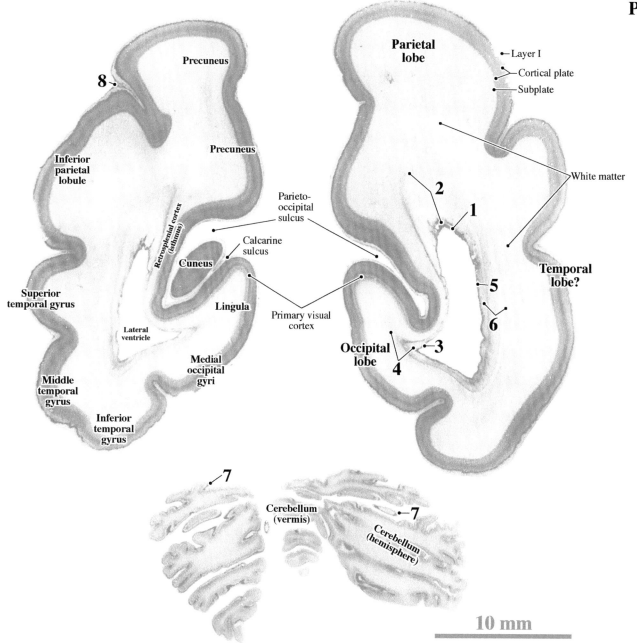

Precuneus

8

Precuneus

Inferior
parietal
lobule

Parietal
lobe

Layer I

Cortical plate

Subplate

White matter

2

1

Parieto-
occipital
sulcus

Calcarine
sulcus

Retrosplenial cortex
(isthmus)

Cuneus

Superior
temporal gyrus

Lingula

Medial
occipital
gyri

Primary visual
cortex

Lateral
ventricle

Temporal
lobe?

5

6

Occipital
lobe

3

4

Middle
temporal
gyrus

Inferior
temporal
gyrus

7

Cerebellum
(vermis)

7

Cerebellum
(hemisphere)

10 mm

PLATE 36A
CR 260 mm
GW 30, Y14-59
Frontal
Section 1621

See detail of the primary visual cortex
in Section 1681 in Plate 40.

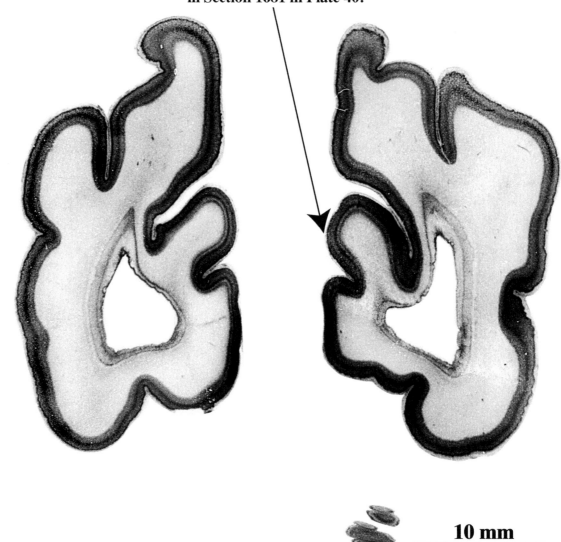

Remnants of the germinal matrix,
migratory streams, and
transitional fields

1 *Parietal NEP and SVZ*

2 *Parietal STF*

3 *Occipital NEP and SVZ*

4 *Occipital STF*

5 *External germinal layer (cerebellum)*

6 *Subpial granular layer (cortical)*

NEP - Neuroepithelium
STF - Stratified transitional field
SVZ - Subventricular zone

10 mm

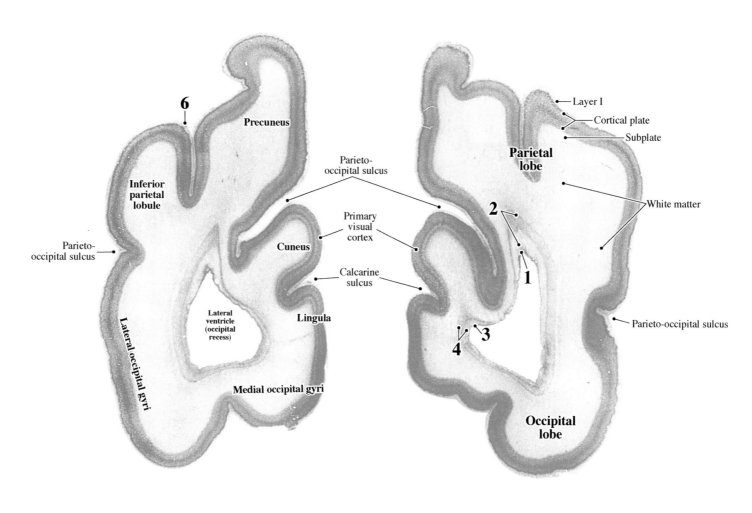

6

Precuneus

Layer I

Cortical plate

Subplate

Parietal lobe

Parieto-occipital sulcus

Inferior parietal lobule

Primary visual cortex

2

White matter

Parieto-occipital sulcus

Cuneus

1

Calcarine sulcus

Parieto-occipital sulcus

Lateral occipital gyri

Lingula

Lateral ventricle (occipital recess)

3

4

Medial occipital gyri

Occipital lobe

Cerebellum (hemisphere)

5

PLATE 37
CR 260 mm
GW 30, Y14-59
Frontal
Section 861
SUPERIOR FRONTAL
GYRUS

See the entire Section 861
in Plates 24A and B.

1.5 mm

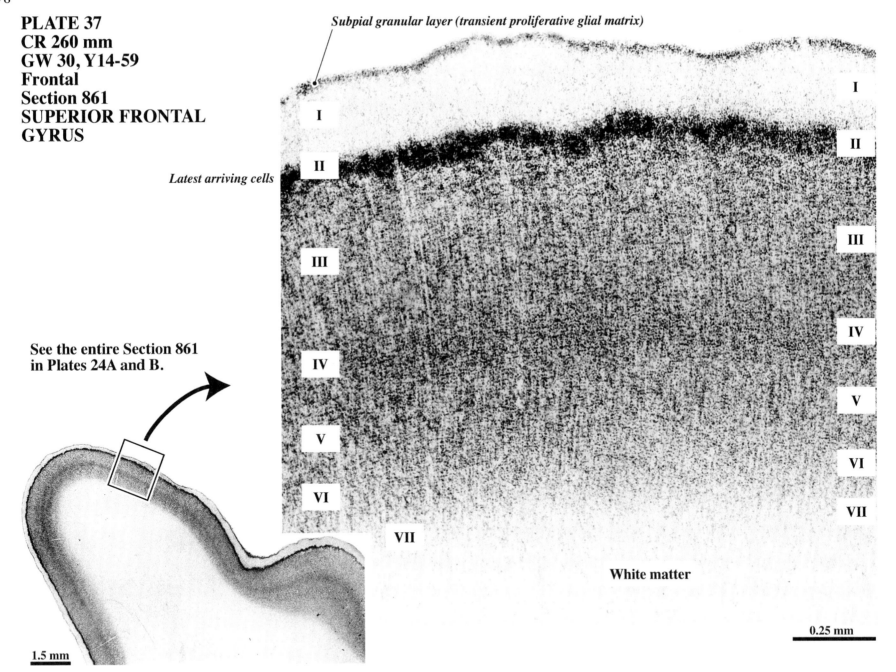

Subpial granular layer (transient proliferative glial matrix)

Latest arriving cells

I

II

III

IV

V

VI

VII

White matter

0.25 mm

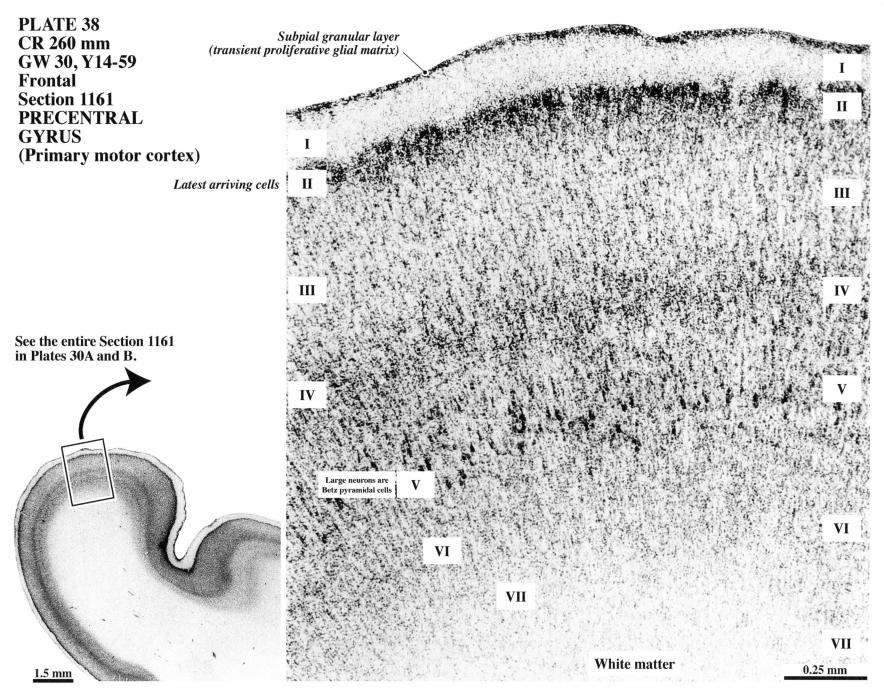

PLATE 38
CR 260 mm
GW 30, Y14-59
Frontal
Section 1161
PRECENTRAL
GYRUS
(Primary motor cortex)

Subpial granular layer
(transient proliferative glial matrix)

I

II

Latest arriving cells

III

See the entire Section 1161
in Plates 30A and B.

IV

Large neurons are
Betz pyramidal cells

V

VI

VII

White matter

I

II

III

IV

V

VI

VII

1.5 mm

0.25 mm

**PLATE 39
CR 260 mm
GW 30, Y14-59
Frontal
Section 1301
SUPERIOR
PARIETAL
LOBULE**

Latest arriving cells

**See the entire Section 1301
in Plates 33A amd B.**

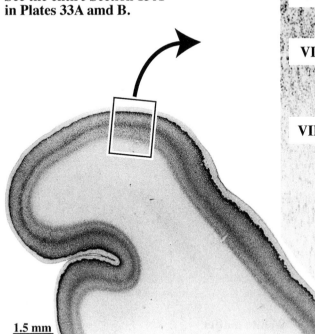

1.5 mm

I

II

III

IV

V

VI

VII

I

II

III

IV

V

VI

VII

White matter

0.25 mm

PLATE 40
CR 260 mm, GW 30, Y14-59, Frontal, Section 1681
CUNEUS (Primary visual cortex)

See the complete nearby Section 1621 in Plates 36A and B.

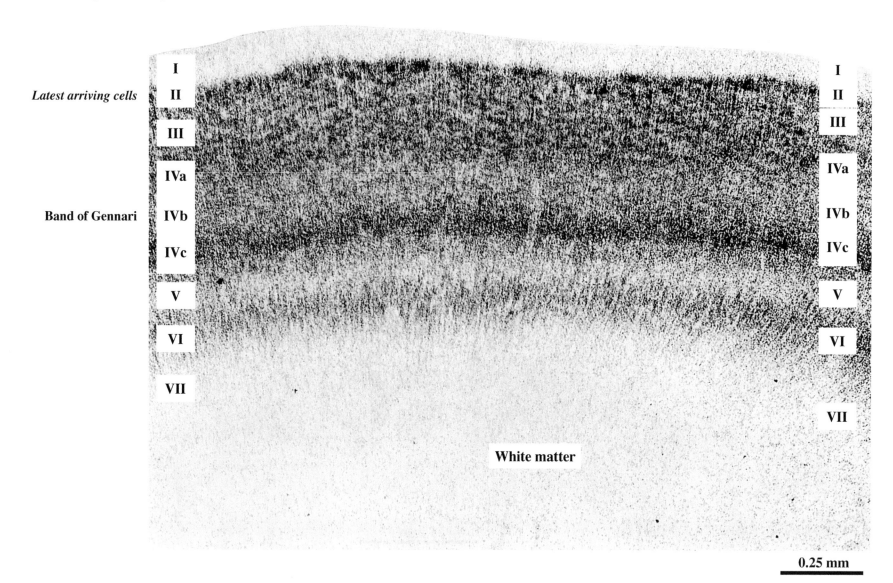

Latest arriving cells

Band of Gennari

I
II
III
IVa
IVb
IVc
V
VI
VII

I
II
III
IVa
IVb
IVc
V
VI
VII

White matter

0.25 mm

PLATE 41A
CR 260 mm, GW 30, Y14-59, Frontal, Section 701

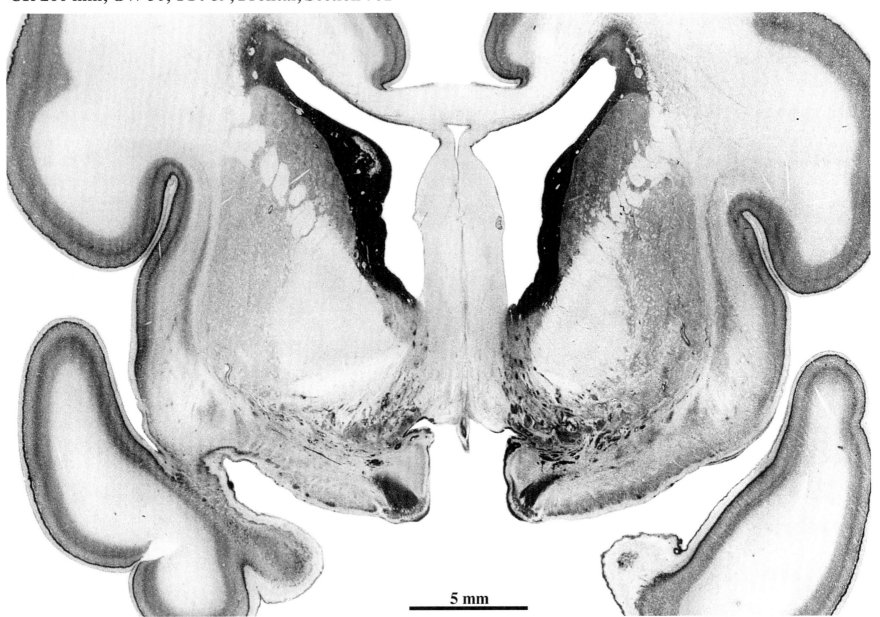

5 mm

See the entire Section 701 in Plates 20A and B.

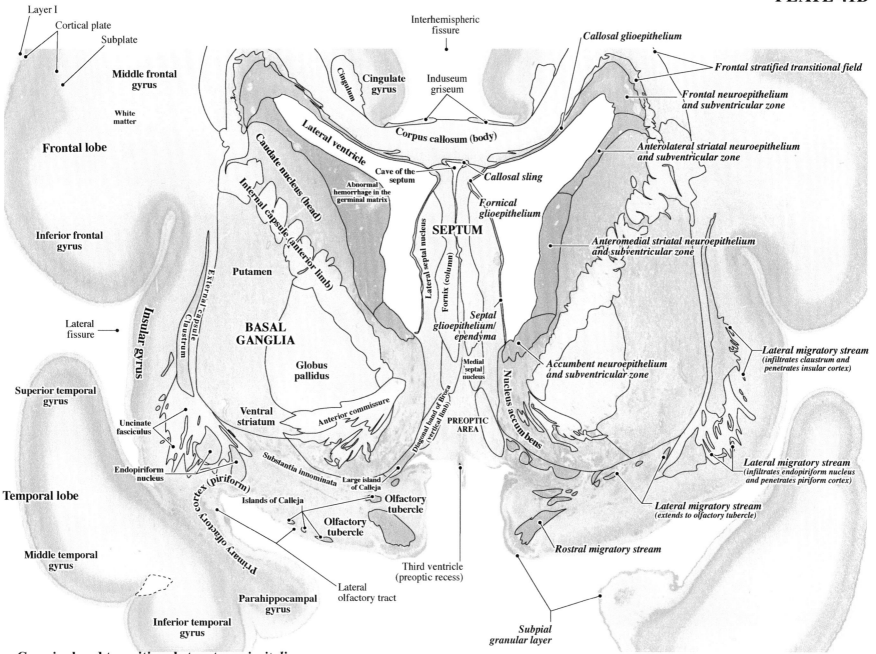

Layer I

Cortical plate

Subplate

Middle frontal gyrus

White matter

Frontal lobe

Inferior frontal gyrus

Interhemispheric fissure

Callosal glioepithelium

Frontal stratified transitional field

Cingulum

Cingulate gyrus

Induseum griseum

Frontal neuroepithelium and subventricular zone

Lateral ventricle

Corpus callosum (body)

Anterolateral striatal neuroepithelium and subventricular zone

Caudate nucleus (head)

Cave of the septum

Callosal sling

Abnormal hemorrhage in the germinal matrix

SEPTUM

Internal capsule (anterior limb)

Fornical glioepithelium

Putamen

Lateral septal nucleus

Anteromedial striatal neuroepithelium and subventricular zone

Lateral fissure

External capsule

Claustrum

BASAL GANGLIA

Fornix (column)

Septal glioepithelium/ ependyma

Accumbent neuroepithelium and subventricular zone

Insular gyrus

Globus pallidus

Medial septal nucleus

Nucleus accumbens

Lateral migratory stream (infiltrates claustrum and penetrates insular cortex)

Superior temporal gyrus

Ventral striatum

Anterior commissure

Diagonal band of Broca (vertical limb)

PREOPTIC AREA

Uncinate fasciculus

Substantia innominata

Endopiriform nucleus

Large island of Calleja

Lateral migratory stream (infiltrates endopiriform nucleus and penetrates piriform cortex)

Temporal lobe

Primary olfactory cortex (piriform)

Islands of Calleja

Olfactory tubercle

Olfactory tubercle

Lateral migratory stream (extends to olfactory tubercle)

Middle temporal gyrus

Lateral olfactory tract

Third ventricle (preoptic recess)

Rostral migratory stream

Parahippocampal gyrus

Inferior temporal gyrus

Subpial granular layer

Germinal and transitional structures in *italics*

PLATE 42A
CR 260 mm, GW 30, Y14-59, Frontal, Section 741

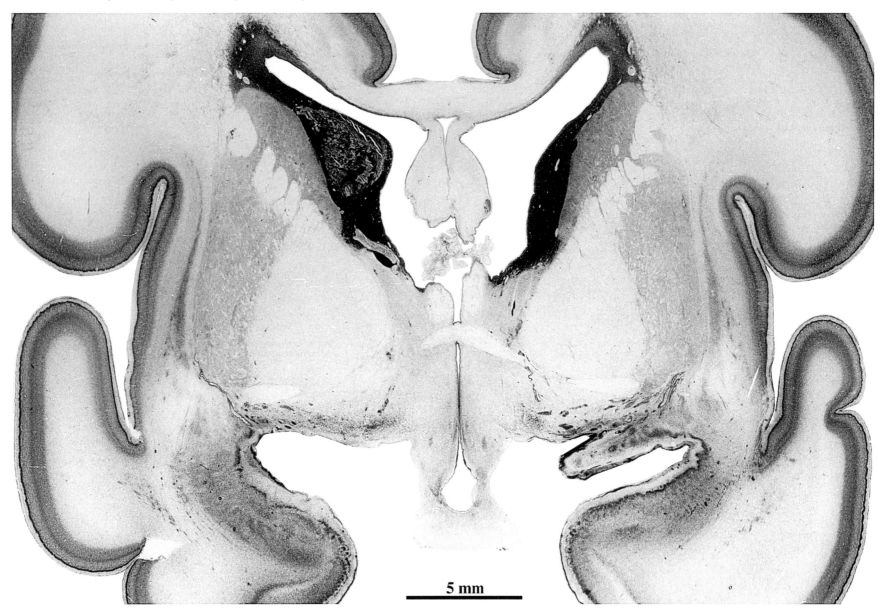

5 mm

See the entire Section 741 in Plates 21A and B.

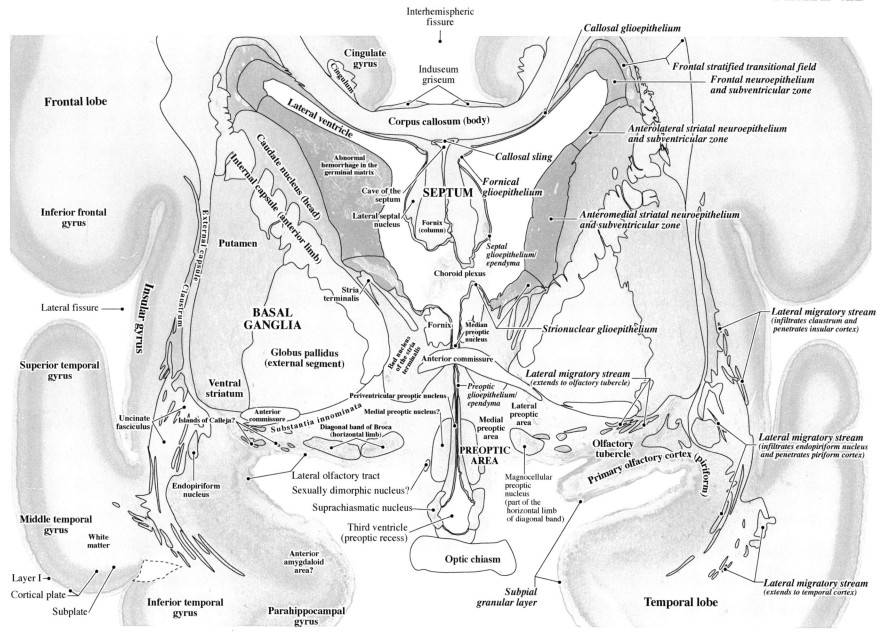

Interhemispheric fissure

Callosal glioepithelium

Cingulate gyrus

Cingulum

Induseum griseum

Frontal stratified transitional field
Frontal neuroepithelium and subventricular zone

Frontal lobe

Lateral ventricle

Corpus callosum (body)

Callosal sling

Anterolateral striatal neuroepithelium and subventricular zone

Inferior frontal gyrus

Caudate nucleus (head)

Internal capsule (anterior limb)

Abnormal hemorrhage in the germinal matrix

Cave of the septum

SEPTUM

Fornical glioepithelium

Anteromedial striatal neuroepithelium and subventricular zone

Lateral septal nucleus

Fornix (column)

Putamen

External capsule

Septal glioepithelium/ ependyma

Choroid plexus

Claustrum

Lateral fissure

Insular gyrus

Stria terminalis

BASAL GANGLIA

Fornix

Median preoptic nucleus

Strionuclear glioepithelium

Lateral migratory stream (infiltrates claustrum and penetrates insular cortex)

Superior temporal gyrus

Globus pallidus (external segment)

Bed nucleus of the stria terminalis

Anterior commissure

Ventral striatum

Periventricular preoptic nucleus

Preoptic glioepithelium/ ependyma

Lateral migratory stream (extends to olfactory tubercle)

Anterior commissure

Medial preoptic nucleus?

Lateral preoptic area

Uncinate fasciculus

Islands of Calleja?

Substantia innominata

Diagonal band of Broca (horizontal limb)

Medial preoptic area

Olfactory tubercle

Lateral migratory stream (infiltrates endopiriform nucleus and penetrates piriform cortex)

PREOPTIC AREA

Endopiriform nucleus

Lateral olfactory tract

Sexually dimorphic nucleus?

Magnocellular preoptic nucleus (part of the horizontal limb of diagonal band)

Primary olfactory cortex (piriform)

Middle temporal gyrus

White matter

Suprachiasmatic nucleus

Third ventricle (preoptic recess)

Layer I

Anterior amygdaloid area?

Cortical plate

Optic chiasm

Lateral migratory stream (extends to temporal cortex)

Subplate

Subpial granular layer

Temporal lobe

Inferior temporal gyrus

Parahippocampal gyrus

Germinal and transitional structures in *italics*

PLATE 43A
CR 260 mm, GW 30, Y14-59, Frontal, Section 781

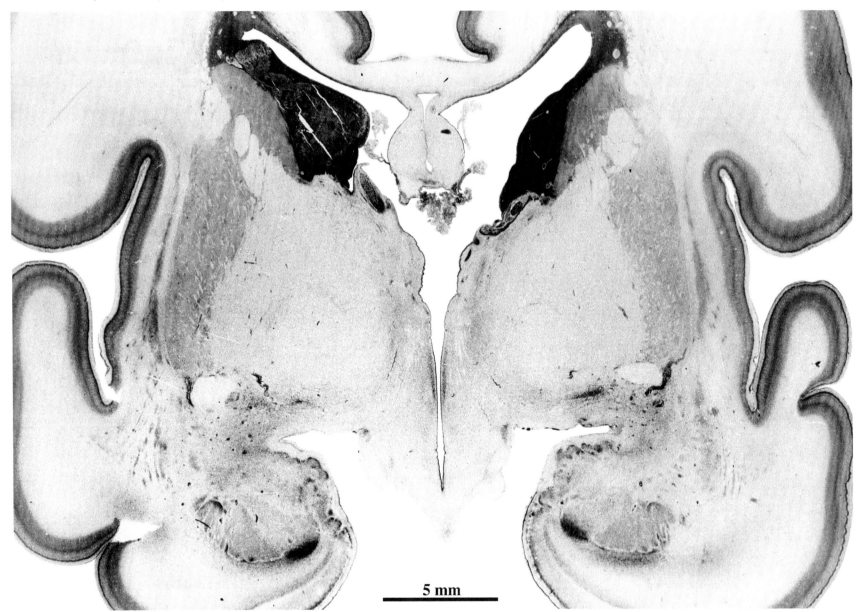

5 mm

See the entire Section 741 in Plates 22A and B.

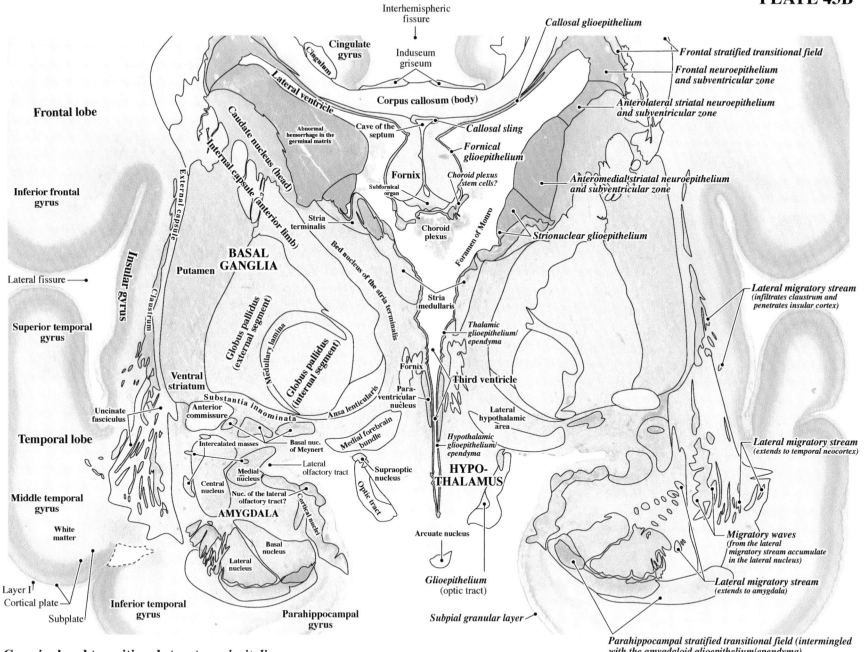

Interhemispheric fissure

Cingulate gyrus

Cingulum

Lateral ventricle

Induseum griseum

Callosal glioepithelium

Frontal stratified transitional field

Frontal neuroepithelium and subventricular zone

Corpus callosum (body)

Frontal lobe

Caudate nucleus (head)

Abnormal hemorrhage in the germinal matrix

Cave of the septum

Callosal sling

Anterolateral striatal neuroepithelium and subventricular zone

Internal capsule (anterior limb)

Fornical glioepithelium

Inferior frontal gyrus

External capsule

Fornix

Choroid plexus stem cells?

Subfornical organ

Anteromedial striatal neuroepithelium and subventricular zone

Stria terminalis

Choroid plexus

Foramen of Monro

Strionuclear glioepithelium

BASAL GANGLIA

Claustrum

Insular gyrus

Bed nucleus of the stria terminalis

Putamen

Lateral fissure

Stria medullaris

Lateral migratory stream (infiltrates claustrum and penetrates insular cortex)

Superior temporal gyrus

Globus pallidus (external segment)

Thalamic glioepithelium/ ependyma

Medullary lamina

Globus pallidus (internal segment)

Fornix

Third ventricle

Ventral striatum

Ansa lenticularis

Para-ventricular nucleus

Lateral hypothalamic area

Lateral migratory stream (extends to temporal neocortex)

Substantia innominata

Anterior commissure

Basal nuc. of Meynert

Medial forebrain bundle

Uncinate fasciculus

Hypothalamic glioepithelium/ ependyma

Temporal lobe

Intercalated masses

Lateral olfactory tract

Supraoptic nucleus

Medial nucleus

Optic tract

HYPO-THALAMUS

Central nucleus

Nuc. of the lateral olfactory tract?

Middle temporal gyrus

Cortical nuclei

AMYGDALA

Migratory waves (from the lateral migratory stream accumulate in the lateral nucleus)

White matter

Basal nucleus

Arcuate nucleus

Lateral migratory stream (extends to amygdala)

Lateral nucleus

Layer I

Cortical plate

Glioepithelium (optic tract)

Subplate

Inferior temporal gyrus

Parahippocampal gyrus

Subpial granular layer

Parahippocampal stratified transitional field (intermingled with the amygdaloid glioepithelium/ependyma)

Germinal and transitional structures in *italics*

PLATE 44A
CR 260 mm, GW 30, Y14-59, Frontal, Section 821

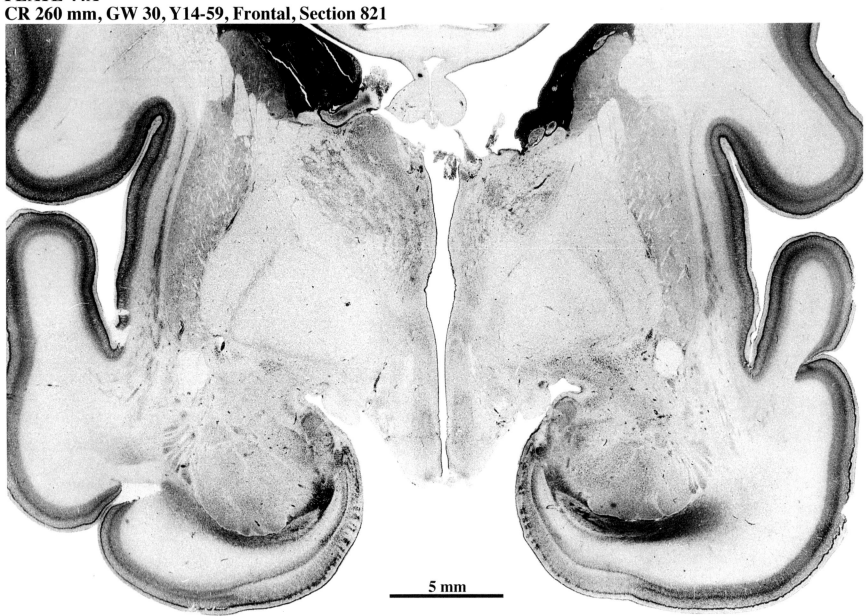

5 mm

See the entire Section 821 in Plates 23A and B.

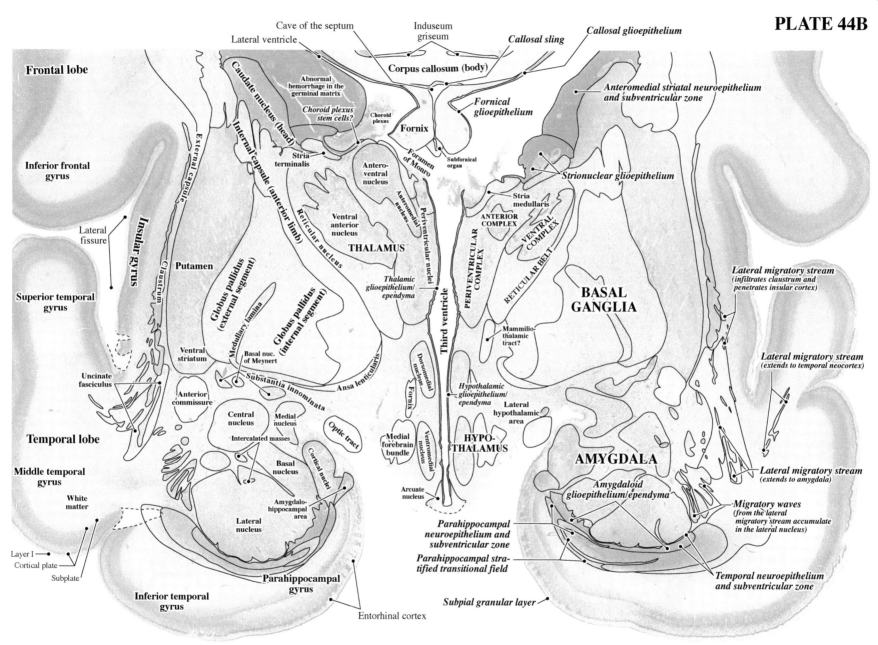

Germinal and transitional structures in *italics*

PLATE 45A
CR 260 mm, GW 30, Y14-59, Frontal, Section 861

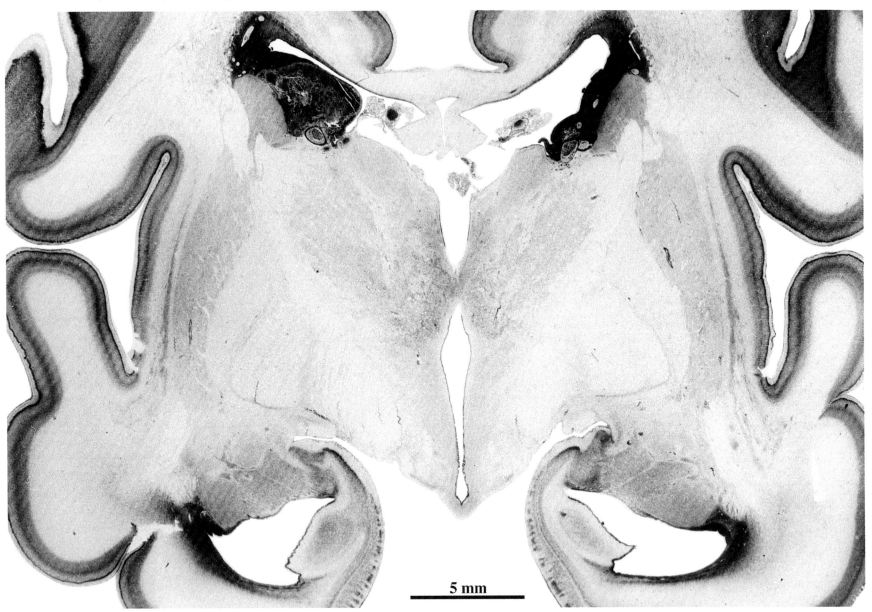

5 mm

See the entire Section 861 in Plates 24A and B.

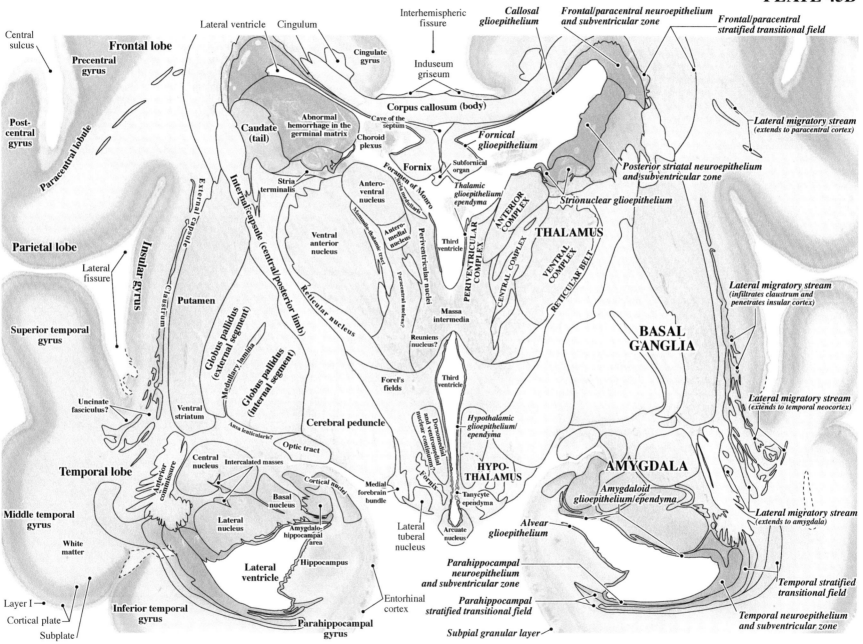

Germinal and transitional structures in *italics*

PLATE 46A
CR 260 mm, GW 30, Y14-59, Frontal, Section 901

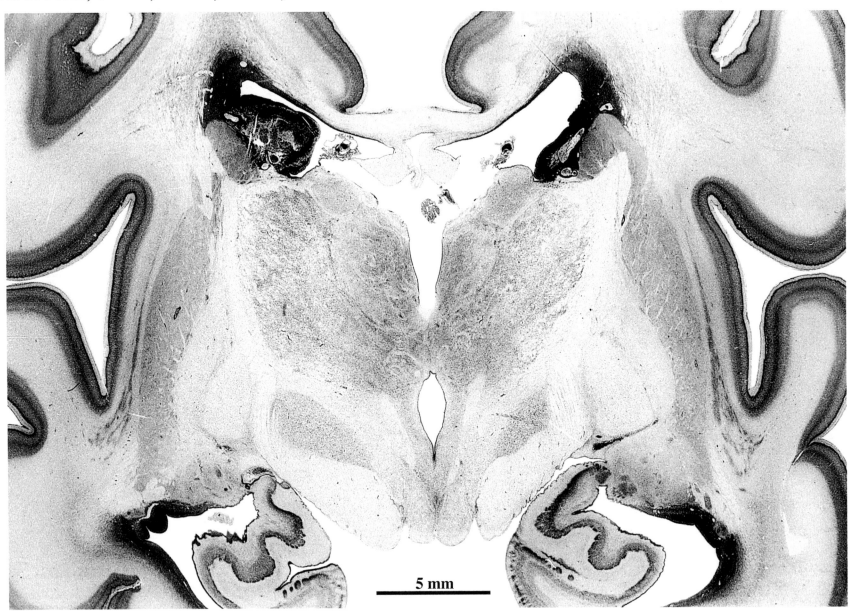

5 mm

See the entire Section 901 in Plates 25A and B.

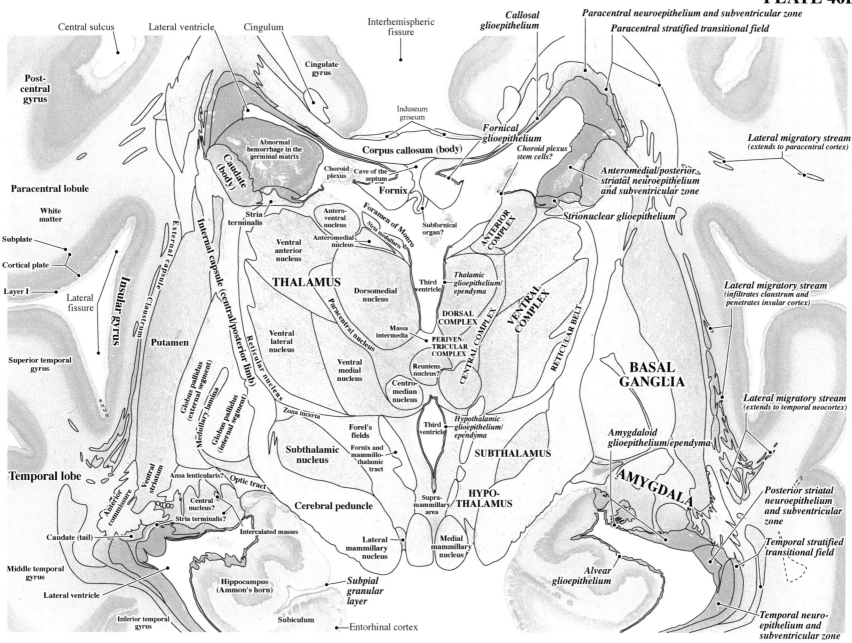

Paracentral neuroepithelium and subventricular zone
Paracentral stratified transitional field

Central sulcus
Lateral ventricle
Cingulum
Interhemispheric fissure
Callosal glioepithelium

Post-central gyrus

Abnormal hemorrhage in the germinal matrix

Induseum griseum

Fornical glioepithelium

Corpus callosum (body)

Choroid plexus stem cells?

Lateral migratory stream
(extends to paracentral cortex)

Paracentral lobule

Caudate (body)

Choroid plexus

Cave of the septum

Fornix

Anteromedial/posterior striatal neuroepithelium and subventricular zone

White matter

Stria terminalis

Foramen of Monro

Subfornical organ?

Strionuclear glioepithelium

Subplate

Antero-ventral nucleus

Cortical plate

Anteromedial nucleus

Stria medullaris

ANTERIOR COMPLEX

Layer I

Ventral anterior nucleus

Lateral fissure

Thalamic glioepithelium/ ependyma

Lateral migratory stream
(infiltrates claustrum and penetrates insular cortex)

Insular gyrus

THALAMUS

Dorsomedial nucleus

Third ventricle

DORSAL COMPLEX

VENTRAL COMPLEX

Superior temporal gyrus

Putamen

Paracentral nucleus

PERIVEN-TRICULAR COMPLEX

CENTRAL COMPLEX

RETICULAR BELT

BASAL GANGLIA

Ventral lateral nucleus

Massa intermedia

Ventral medial nucleus

Reuniens nucleus?

Centro-median nucleus

Globus pallidus (external segment)

Medullary lamina

Globus pallidus (internal segment)

Zona incerta

Forel's fields

Third ventricle

Hypothalamic glioepithelium/ ependyma

Amygdaloid glioepithelium/ependyma

Lateral migratory stream
(extends to temporal neocortex)

Temporal lobe

Subthalamic nucleus

Fornix and mammillo-thalamic tract

SUBTHALAMUS

AMYGDALA

Ventral striatum

Ansa lenticularis?

Optic tract

Cerebral peduncle

Posterior striatal neuroepithelium and subventricular zone

Anterior commissure

Central nucleus?

Supra-mammillary area

HYPO-THALAMUS

Caudate (tail)

Stria terminalis?

Intercalated masses

Temporal stratified transitional field

Middle temporal gyrus

Lateral ventricle

Lateral mammillary nucleus

Medial mammillary nucleus

Alvear glioepithelium

Hippocampus (Ammon's horn)

Subpial granular layer

Inferior temporal gyrus

Subiculum

Entorhinal cortex

Temporal neuro-epithelium and subventricular zone

Germinal and transitional structures in *italics*

94

PLATE 47A
CR 260 mm
GW 30
Y14-59
Frontal
Section 981

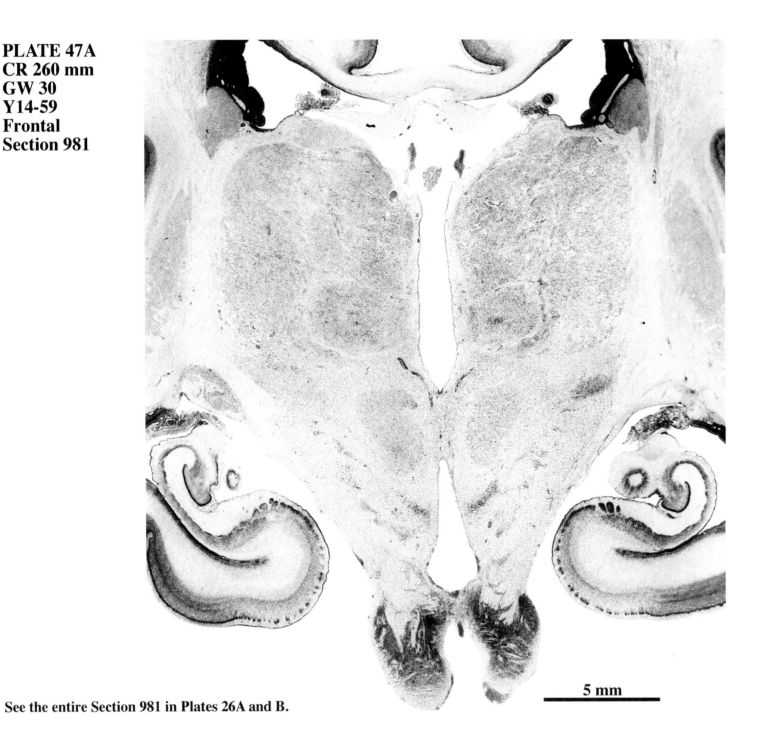

5 mm

See the entire Section 981 in Plates 26A and B.

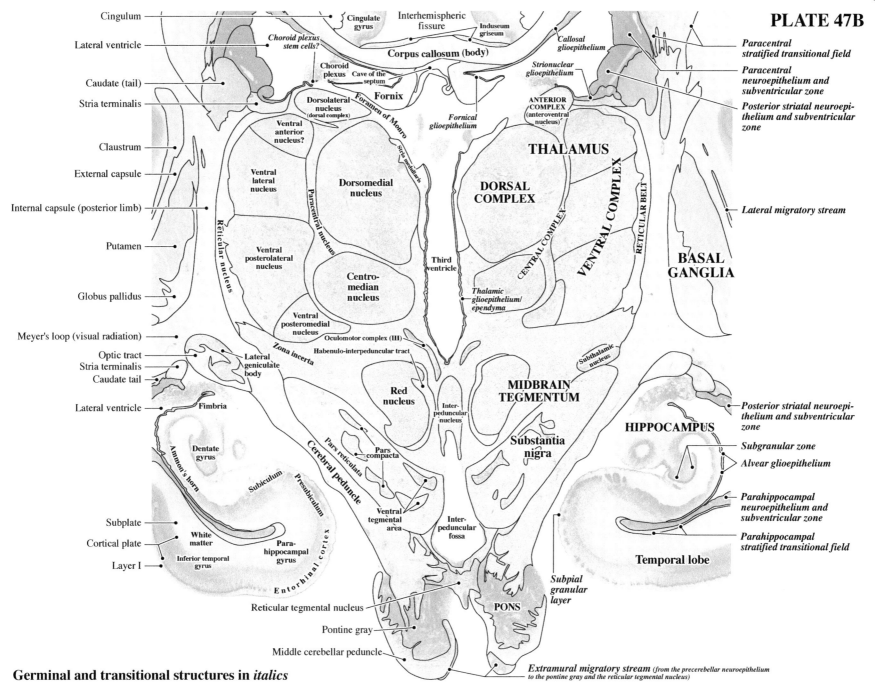

Cingulum

Lateral ventricle

Caudate (tail)

Stria terminalis

Claustrum

External capsule

Internal capsule (posterior limb)

Putamen

Globus pallidus

Meyer's loop (visual radiation)

Optic tract

Stria terminalis

Caudate tail

Lateral ventricle

Subplate

Cortical plate

Layer I

Cingulate gyrus

Interhemispheric fissure

Induseum griseum

Choroid plexus stem cells?

Corpus callosum (body)

Choroid plexus

Strionuclear glioepithelium

Cave of the septum

Callosal glioepithelium

Fornix

Foramen of Monro

Dorsolateral nucleus *(dorsal complex)*

Ventral anterior nucleus?

Fornical glioepithelium

ANTERIOR COMPLEX *(anteroventral nucleus)*

Stria medullaris

Ventral lateral nucleus

Dorsomedial nucleus

Paracentral nucleus

DORSAL COMPLEX

THALAMUS

CENTRAL COMPLEX

VENTRAL COMPLEX

RETICULAR BELT

Reticular nucleus

Ventral posterolateral nucleus

Centro-median nucleus

Third ventricle

Ventral posteromedial nucleus

Thalamic glioepithelium/ ependyma

Oculomotor complex (III)

Zona incerta

Habenulo-interpeduncular tract

Subthalamic nucleus

BASAL GANGLIA

Lateral geniculate body

Red nucleus

Inter-peduncular nucleus

MIDBRAIN TEGMENTUM

Fimbria

Dentate gyrus

Ammon's horn

Subiculum

Presubiculum

Pars reticulata

Cerebral peduncle

Pars compacta

Substantia nigra

HIPPOCAMPUS

White matter

Para-hippocampal gyrus

Ventral tegmental area

Inter-peduncular fossa

Inferior temporal gyrus

Entorhinal cortex

Temporal lobe

Reticular tegmental nucleus

Pontine gray

Middle cerebellar peduncle

Subpial granular layer

PONS

Paracentral stratified transitional field

Paracentral neuroepithelium and subventricular zone

Posterior striatal neuroepi-thelium and subventricular zone

Lateral migratory stream

Posterior striatal neuroepi-thelium and subventricular zone

Subgranular zone

Alvear glioepithelium

Parahippocampal neuroepithelium and subventricular zone

Parahippocampal stratified transitional field

Germinal and transitional structures in *italics*

Extramural migratory stream (from the precerebellar neuroepithelium to the pontine gray and the reticular tegmental nucleus)

PLATE 48A
CR 260 mm
GW 30
Y14-59
Frontal
Section 1021

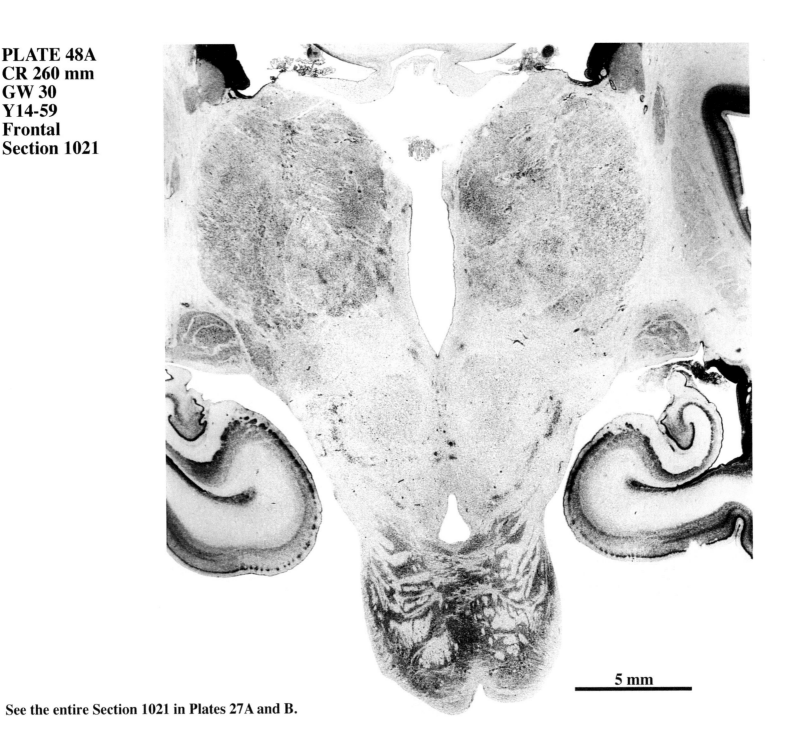

5 mm

See the entire Section 1021 in Plates 27A and B.

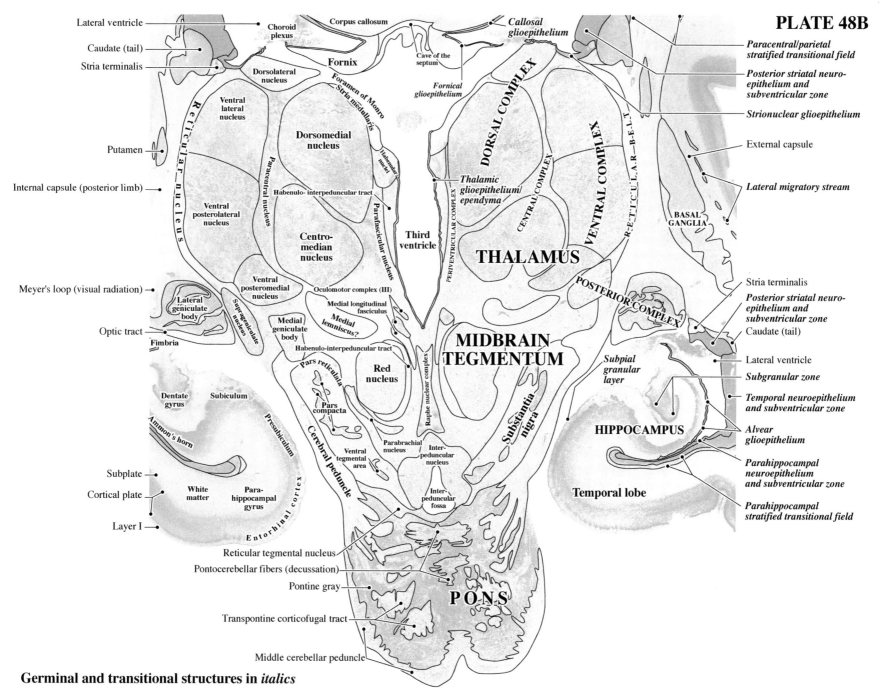

PLATE 48B

Lateral ventricle

Caudate (tail)

Stria terminalis

Choroid plexus

Corpus callosum

Callosal glioepithelium

Paracentral/parietal stratified transitional field

Posterior striatal neuro-epithelium and subventricular zone

Fornix

Cave of the septum

Fornical glioepithelium

Dorsolateral nucleus

Foramen of Monro

Stria medullaris

DORSAL COMPLEX

Strionuclear glioepithelium

Ventral lateral nucleus

Dorsomedial nucleus

Habenular nuclei

Putamen

Paracentral nucleus

Habenulo- interpeduncular tract

Thalamic glioepithelium/ ependyma

CENTRAL COMPLEX

VENTRAL COMPLEX

R-E-T-I-C-U-L-A-R—B-E-L-T

External capsule

Internal capsule (posterior limb)

Ventral posterolateral nucleus

Parafascicular nucleus

Third ventricle

PERIVENTRICULAR COMPLEX

Lateral migratory stream

BASAL GANGLIA

Centro-median nucleus

THALAMUS

Meyer's loop (visual radiation)

Ventral posteromedial nucleus

Oculomotor complex (III)

POSTERIOR COMPLEX

Stria terminalis

Lateral geniculate body

Suprageniculate nucleus

Medial longitudinal fasciculus

Posterior striatal neuro-epithelium and subventricular zone

Optic tract

Medial geniculate body

Medial lemniscus?

MIDBRAIN TEGMENTUM

Caudate (tail)

Fimbria

Habenulo-interpeduncular tract

Raphe nuclear complex

Subpial granular layer

Lateral ventricle

Pars reticulata

Red nucleus

Subgranular zone

Dentate gyrus

Subiculum

Ammon's horn

Presubiculum

Pars compacta

Substantia nigra

HIPPOCAMPUS

Temporal neuroepithelium and subventricular zone

Alvear glioepithelium

Subplate

White matter

Para-hippocampal gyrus

Cerebral peduncle

Ventral tegmental area

Parabrachial nucleus

Inter-peduncular nucleus

Temporal lobe

Parahippocampal neuroepithelium and subventricular zone

Cortical plate

Layer I

Entorhinal cortex

Inter-peduncular fossa

Parahippocampal stratified transitional field

Reticular tegmental nucleus

Pontocerebellar fibers (decussation)

Pontine gray

P O N S

Transpontine corticofugal tract

Middle cerebellar peduncle

Reticular nucleus

Germinal and transitional structures in *italics*

PLATE 49A
CR 260 mm
GW 30
Y14-59
Frontal
Section 1061

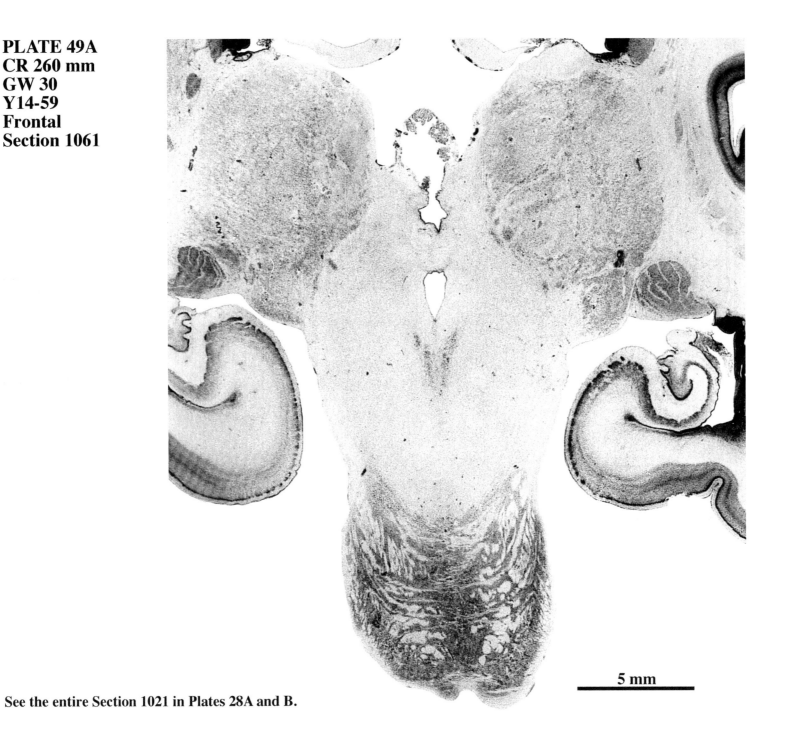

5 mm

See the entire Section 1021 in Plates 28A and B.

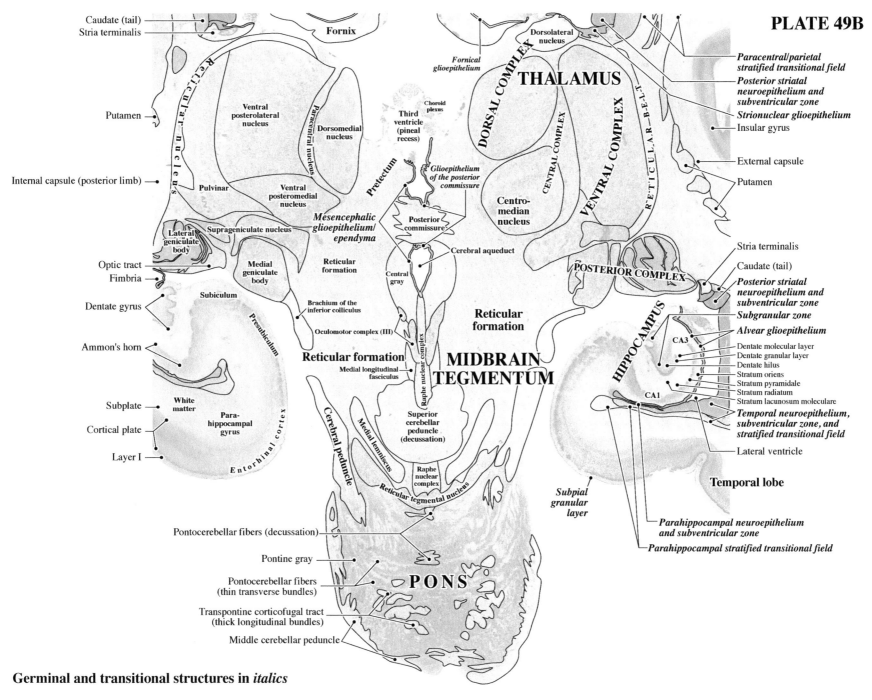

Caudate (tail)
Stria terminalis
Fornix
Dorsolateral nucleus
Paracentral/parietal stratified transitional field
Posterior striatal neuroepithelium and subventricular zone
Fornical glioepithelium
THALAMUS
DORSAL COMPLEX
Strionuclear glioepithelium
Putamen
Ventral posterolateral nucleus
Paracentral nucleus
Dorsomedial nucleus
Choroid plexus
Third ventricle (pineal recess)
CENTRAL COMPLEX
VENTRAL COMPLEX
Insular gyrus
External capsule
Internal capsule (posterior limb)
Pulvinar
Ventral posteromedial nucleus
Pretectum
Glioepithelium of the posterior commissure
Centro-median nucleus
RETICULAR BELT
Putamen
Lateral geniculate body
Suprageniculate nucleus
Mesencephalic glioepithelium/ ependyma
Posterior commissure
Cerebral aqueduct
POSTERIOR COMPLEX
Stria terminalis
Caudate (tail)
Optic tract
Medial geniculate body
Reticular formation
Central gray
Posterior striatal neuroepithelium and subventricular zone
Fimbria
Dentate gyrus
Subiculum
Brachium of the inferior colliculus
Reticular formation
Subgranular zone
Alvear glioepithelium
CA3
Ammon's horn
Presubiculum
Oculomotor complex (III)
MIDBRAIN TEGMENTUM
Dentate molecular layer
Dentate granular layer
Dentate hilus
Stratum oriens
Stratum pyramidale
Stratum radiatum
Stratum lacunosum moleculare
Reticular formation
Medial longitudinal fasciculus
HIPPOCAMPUS
Raphe nuclear complex
CA1
Temporal neuroepithelium, subventricular zone, and stratified transitional field
Subplate
White matter
Para-hippocampal gyrus
Entorhinal cortex
Cerebral peduncle
Medial lemniscus
Superior cerebellar peduncle (decussation)
Lateral ventricle
Cortical plate
Layer I
Raphe nuclear complex
Temporal lobe
Subpial granular layer
Reticular tegmental nucleus
Pontocerebellar fibers (decussation)
Parahippocampal neuroepithelium and subventricular zone
Pontine gray
Parahippocampal stratified transitional field
Pontocerebellar fibers (thin transverse bundles)
PONS
Transpontine corticofugal tract (thick longitudinal bundles)
Middle cerebellar peduncle

Germinal and transitional structures in *italics*

PLATE 50A
CR 260 mm
GW 30
Y14-59
Frontal
Section 1101

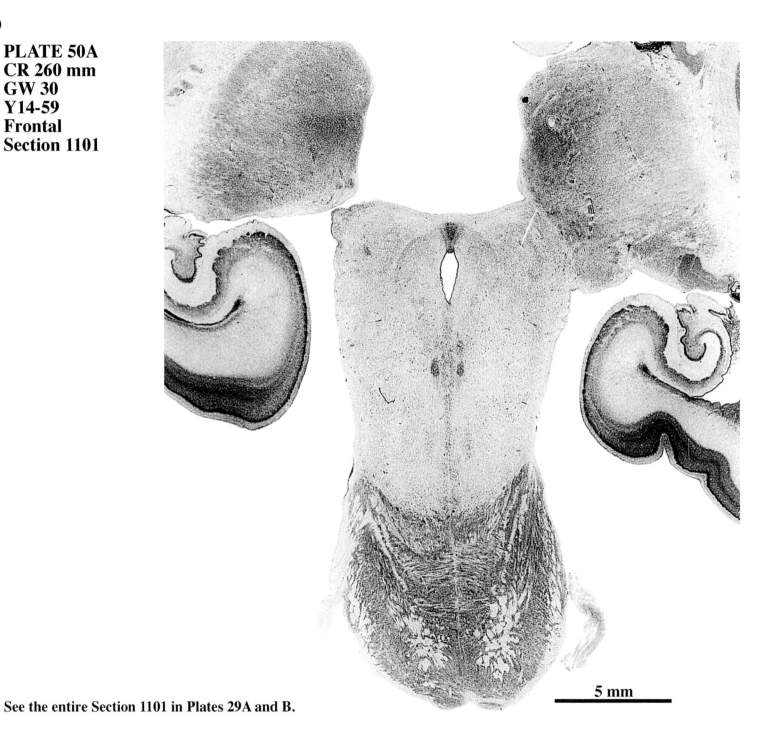

5 mm

See the entire Section 1101 in Plates 29A and B.

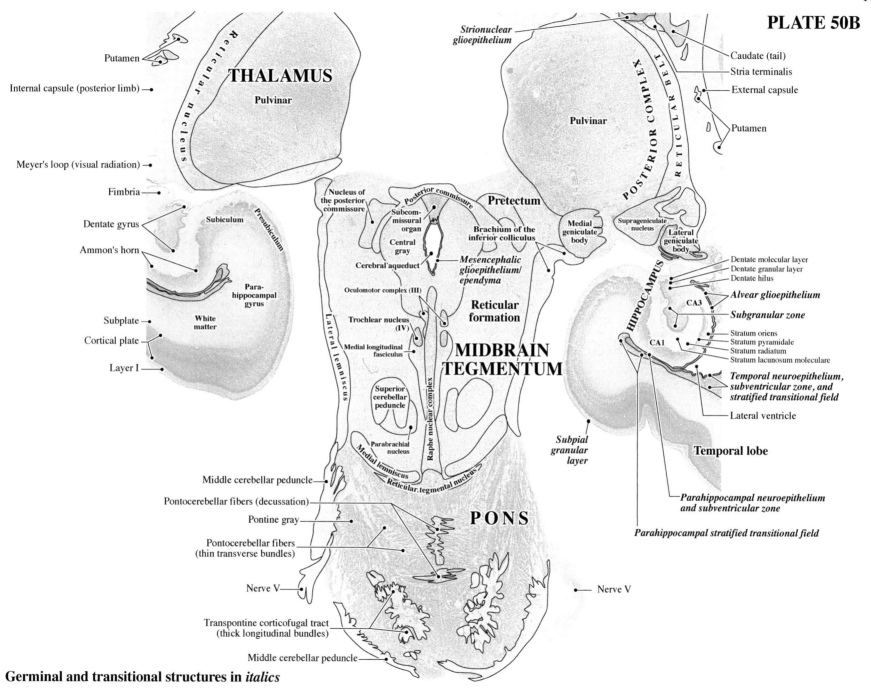

Putamen

Internal capsule (posterior limb)

Meyer's loop (visual radiation)

Fimbria

Dentate gyrus

Ammon's horn

Subplate

Cortical plate

Layer I

THALAMUS

Pulvinar

Reticular nucleus

Subiculum

Presubiculum

Para-hippocampal gyrus

White matter

Strionuclear glioepithelium

Nucleus of the posterior commissure

Subcom-missural organ

Central gray

Cerebral aqueduct

Posterior commissure

Pretectum

Brachium of the inferior colliculus

Mesencephalic glioepithelium/ ependyma

Oculomotor complex (III)

Reticular formation

Trochlear nucleus (IV)

Medial longitudinal fasciculus

Lateral lemniscus

Superior cerebellar peduncle

Parabrachial nucleus

Medial lemniscus

Raphe nuclear complex

Reticular tegmental nucleus

MIDBRAIN TEGMENTUM

Pulvinar

POSTERIOR COMPLEX

RETICULAR BELT

Supra-geniculate nucleus

Medial geniculate body

Lateral geniculate body

HIPPOCAMPUS

Caudate (tail)

Stria terminalis

External capsule

Putamen

Dentate molecular layer

Dentate granular layer

Dentate hilus

Alvear glioepithelium

CA3

Subgranular zone

Stratum oriens

Stratum pyramidale

Stratum radiatum

Stratum lacunosum moleculare

CA1

Temporal neuroepithelium, subventricular zone, and stratified transitional field

Lateral ventricle

Temporal lobe

Middle cerebellar peduncle

Pontocerebellar fibers (decussation)

Pontine gray

Pontocerebellar fibers (thin transverse bundles)

Nerve V

Transpontine corticofugal tract (thick longitudinal bundles)

Middle cerebellar peduncle

PONS

Subpial granular layer

Parahippocampal neuroepithelium and subventricular zone

Parahippocampal stratified transitional field

Nerve V

Germinal and transitional structures in *italics*

PLATE 51A
CR 260 mm
GW 30
Y14-59
Frontal
Section 1161

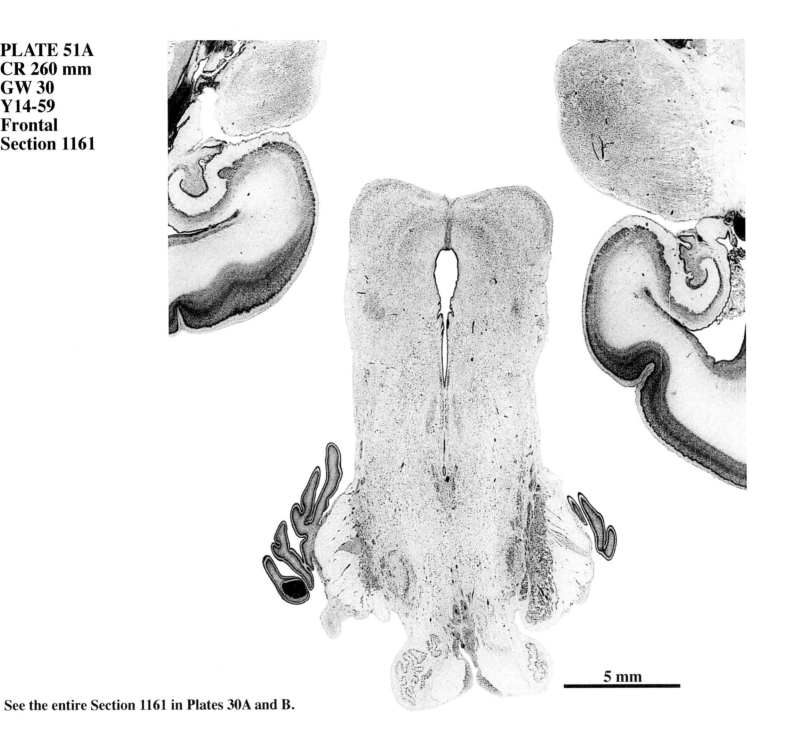

5 mm

See the entire Section 1161 in Plates 30A and B.

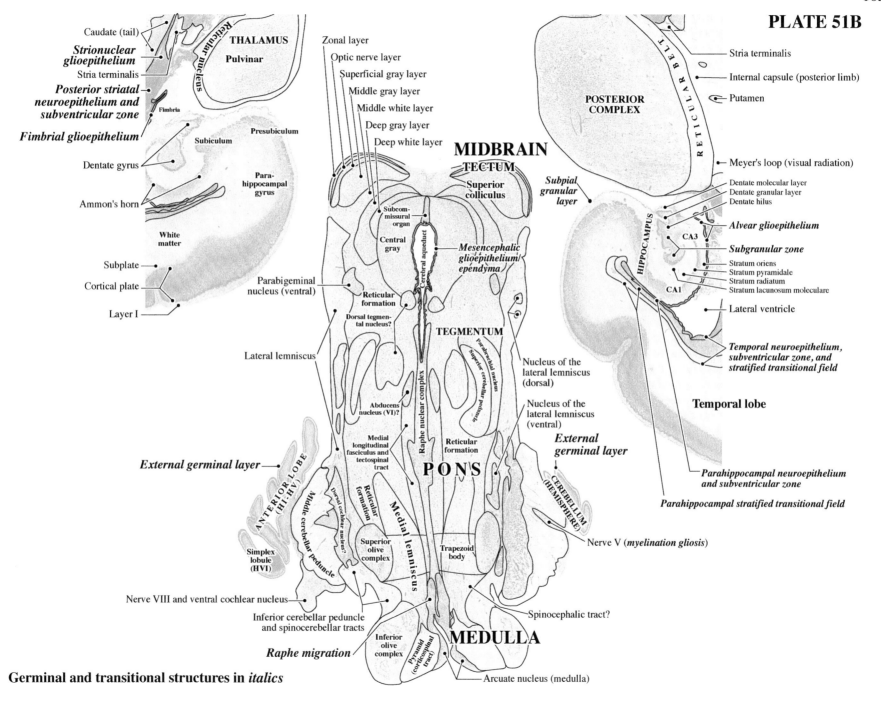

Caudate (tail)

THALAMUS

Strionuclear glioepithelium

Pulvinar

Stria terminalis

Posterior striatal neuroepithelium and subventricular zone

Fimbria

Fimbrial glioepithelium

Presubiculum

Subiculum

Para-hippocampal gyrus

Dentate gyrus

Ammon's horn

White matter

Subplate

Cortical plate

Layer I

Reticular nucleus

Zonal layer

Optic nerve layer

Superficial gray layer

Middle gray layer

Middle white layer

Deep gray layer

Deep white layer

MIDBRAIN

TECTUM

Superior colliculus

Subcommissural organ

Central gray

Cerebral aqueduct

Mesencephalic glioepithelium/ ependyma

Parabigeminal nucleus (ventral)

Reticular formation

Dorsal tegmental nucleus?

TEGMENTUM

Parabrachial nucleus

Superior cerebellar peduncle

Lateral lemniscus

Abducens nucleus (VI)?

Raphe nuclear complex

Medial longitudinal fasciculus and tectospinal tract

PONS

Reticular formation

External germinal layer

ANTERIOR LOBE (HI-HV)

Middle cerebellar peduncle

Dorsal cochlear nucleus?

Reticular formation

Medial lemniscus

Simplex lobule (HVI)

Middle cerebellar peduncle

Superior olive complex

Trapezoid body

Nerve VIII and ventral cochlear nucleus

Inferior cerebellar peduncle and spinocerebellar tracts

Raphe migration

Inferior olive complex

Pyramid (corticospinal tract)

MEDULLA

Arcuate nucleus (medulla)

Stria terminalis

Internal capsule (posterior limb)

Putamen

RETICULAR BELT

POSTERIOR COMPLEX

Meyer's loop (visual radiation)

Dentate molecular layer

Dentate granular layer

Dentate hilus

Subpial granular layer

Alvear glioepithelium

CA3

Subgranular zone

HIPPOCAMPUS

Stratum oriens

Stratum pyramidale

Stratum radiatum

Stratum lacunosum moleculare

CA1

Lateral ventricle

Temporal neuroepithelium, subventricular zone, and stratified transitional field

Nucleus of the lateral lemniscus (dorsal)

Nucleus of the lateral lemniscus (ventral)

External germinal layer

CEREBELLUM (HEMISPHERE)

Temporal lobe

Parahippocampal neuroepithelium and subventricular zone

Parahippocampal stratified transitional field

Nerve V (*myelination gliosis*)

Spinocephalic tract?

Germinal and transitional structures in *italics*

PLATE 52A
CR 260 mm
GW 30
Y14-59
Frontal
Section 1201

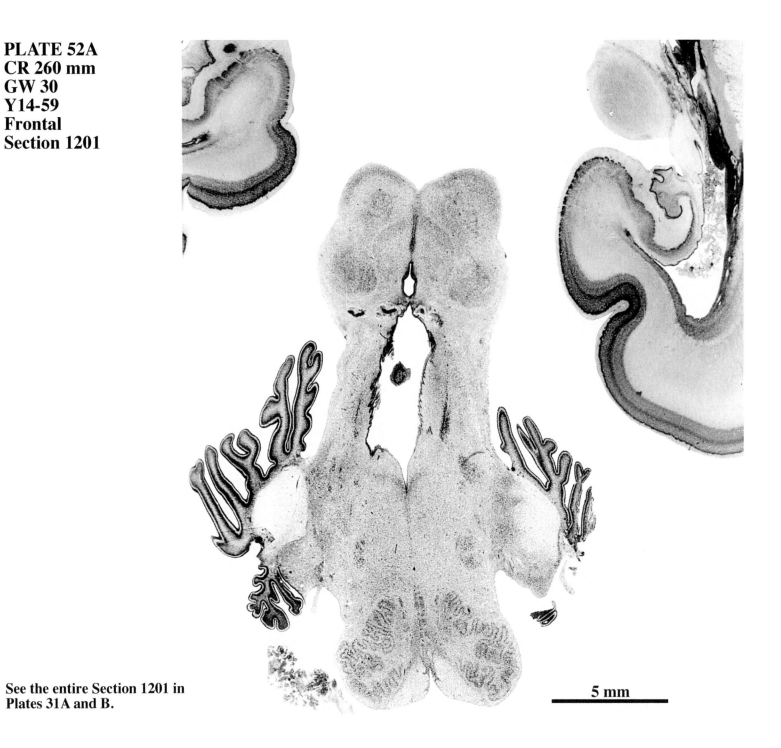

See the entire Section 1201 in
Plates 31A and B.

5 mm

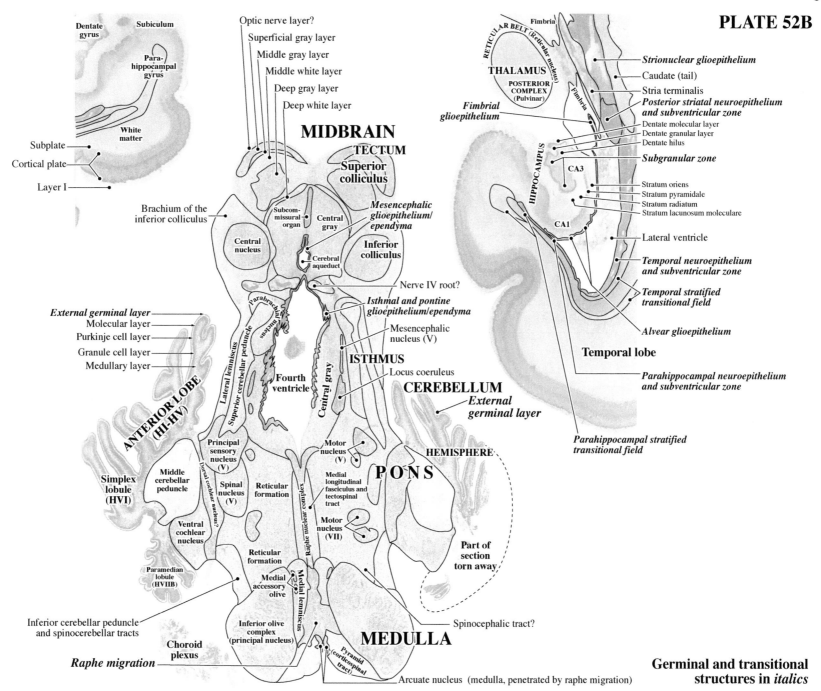

Dentate gyrus

Subiculum

Para-hippocampal gyrus

White matter

Subplate

Cortical plate

Layer I

Optic nerve layer?

Superficial gray layer

Middle gray layer

Middle white layer

Deep gray layer

Deep white layer

MIDBRAIN

TECTUM

Superior colliculus

Brachium of the inferior colliculus

Subcommissural organ

Central gray

Mesencephalic glioepithelium/ ependyma

Central nucleus

Cerebral aqueduct

Inferior colliculus

Nerve IV root?

Isthmal and pontine glioepithelium/ependyma

Mesencephalic nucleus (V)

ISTHMUS

Locus coeruleus

CEREBELLUM

External germinal layer

Molecular layer

Purkinje cell layer

Granule cell layer

Medullary layer

Parabrachial nucleus

Lateral lemniscus

Superior cerebellar peduncle

Fourth ventricle

Central gray

External germinal layer

ANTERIOR LOBE (HI–HV)

Simplex lobule (HVI)

Middle cerebellar peduncle

Dorsal cochlear nucleus?

Principal sensory nucleus (V)

Spinal nucleus (V)

Reticular formation

Raphe nuclear complex

Motor nucleus (V)

Medial longitudinal fasciculus and tectospinal tract

P O N S

HEMISPHERE

Ventral cochlear nucleus

Motor nucleus (VII)

Paramedian lobule (HVIIB)

Reticular formation

Medial accessory olive

Medial lemniscus

Part of section torn away

Inferior cerebellar peduncle and spinocerebellar tracts

Choroid plexus

Inferior olive complex (principal nucleus)

MEDULLA

Spinocephalic tract?

Raphe migration

Pyramid (corticospinal tract)

Arcuate nucleus (medulla, penetrated by raphe migration)

Fimbria

RETICULAR BELT (Reticular nucleus)

THALAMUS

POSTERIOR COMPLEX (Pulvinar)

Fimbrial glioepithelium

Fimbria

HIPPOCAMPUS

CA3

CA1

Strionuclear glioepithelium

Caudate (tail)

Stria terminalis

Posterior striatal neuroepithelium and subventricular zone

Dentate molecular layer

Dentate granular layer

Dentate hilus

Subgranular zone

Stratum oriens

Stratum pyramidale

Stratum radiatum

Stratum lacunosum moleculare

Lateral ventricle

Temporal neuroepithelium and subventricular zone

Temporal stratified transitional field

Alvear glioepithelium

Temporal lobe

Parahippocampal neuroepithelium and subventricular zone

Parahippocampal stratified transitional field

Germinal and transitional structures in *italics*

PLATE 53A
CR 260 mm
GW 30
Y14-59
Frontal
Section 1261

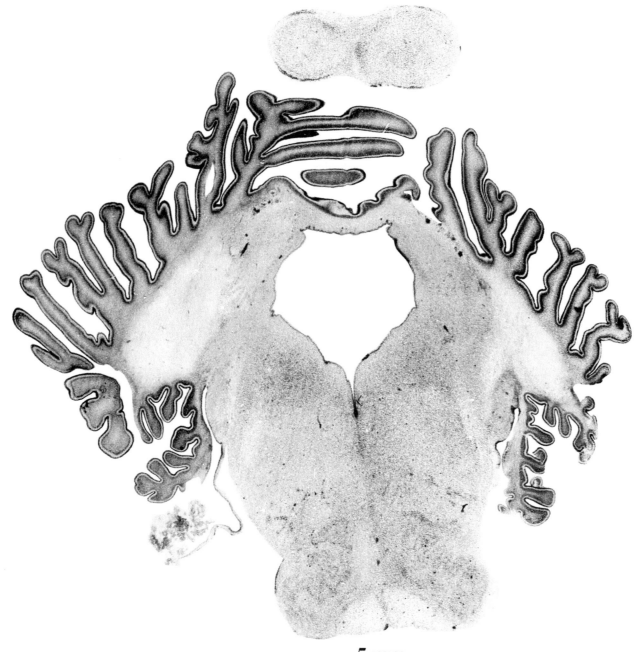

See the entire Section 1261 in Plates 32A and B.

5 mm

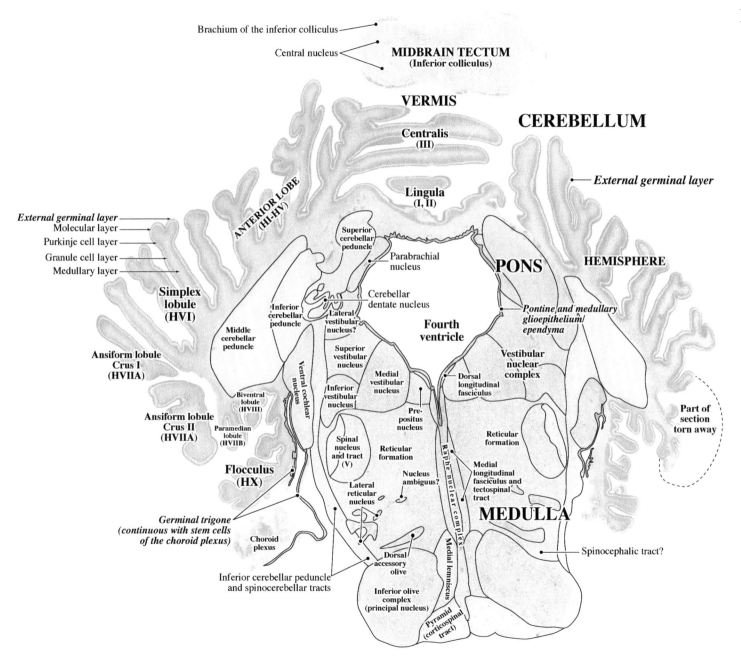

Brachium of the inferior colliculus

Central nucleus

MIDBRAIN TECTUM
(Inferior colliculus)

VERMIS

CEREBELLUM

Centralis
(III)

External germinal layer

Lingula
(I, II)

ANTERIOR LOBE
(HI-HV)

External germinal layer
Molecular layer
Purkinje cell layer
Granule cell layer
Medullary layer

Superior
cerebellar
peduncle

Parabrachial
nucleus

PONS

HEMISPHERE

**Simplex
lobule
(HVI)**

Inferior
cerebellar
peduncle

Cerebellar
dentate nucleus

*Pontine and medullary
glioepithelium/
ependyma*

Lateral
vestibular
nucleus?

**Fourth
ventricle**

Middle
cerebellar
peduncle

Superior
vestibular
nucleus

**Vestibular
nuclear
complex**

**Ansiform lobule
Crus I
(HVIIA)**

Ventral cochlear nucleus

Medial
vestibular
nucleus

Dorsal
longitudinal
fasciculus

Biventral
lobule
(HVIII)

Inferior
vestibular
nucleus

Pre-
positus
nucleus

Reticular
formation

**Part of
section
torn away**

**Ansiform lobule
Crus II
(HVIIA)**

Paramedian
lobule
(HVIIB)

Spinal
nucleus
and tract
(V)

Reticular
formation

Raphe nuclear complex

Medial
longitudinal
fasciculus and
tectospinal
tract

**Flocculus
(HX)**

Nucleus
ambiguus?

Lateral
reticular
nucleus

Medial lemniscus

MEDULLA

Germinal trigone
(continuous with stem cells
of the choroid plexus)

Choroid
plexus

Dorsal
accessory
olive

Spinocephalic tract?

Inferior cerebellar peduncle
and spinocerebellar tracts

Inferior olive
complex
(principal nucleus)

Pyramid
(corticospinal
tract)

Germinal and transitional structures in *italics*

PLATE 54A
CR 260 mm
GW 30
Y14-59
Frontal
Section 1301

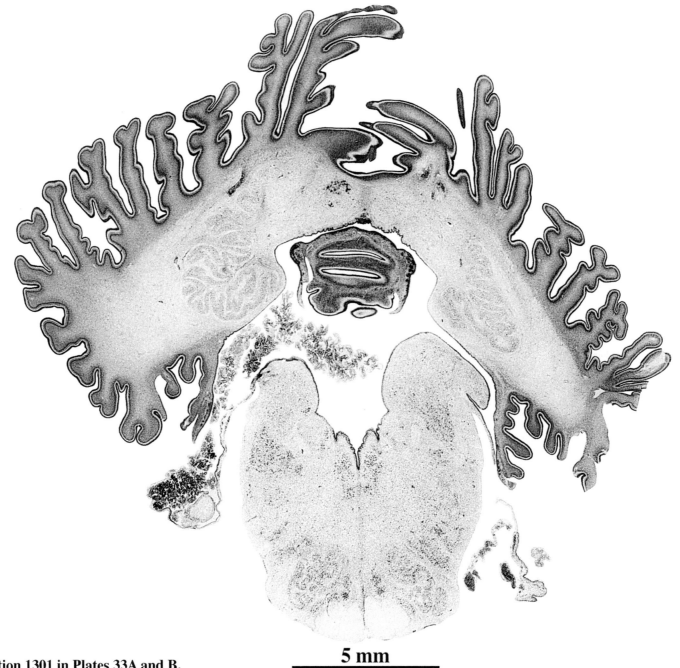

5 mm

See the entire Section 1301 in Plates 33A and B.

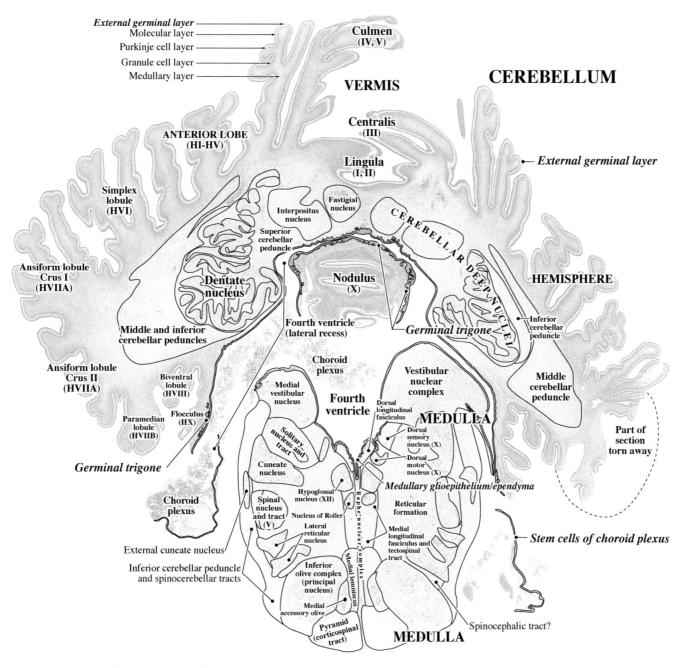

External germinal layer

Molecular layer

Purkinje cell layer

Granule cell layer

Medullary layer

Culmen
(IV, V)

VERMIS

CEREBELLUM

Centralis
(III)

ANTERIOR LOBE
(HI-HV)

Lingula
(I, II)

External germinal layer

Simplex
lobule
(HVI)

Fastigial
nucleus

Interpositus
nucleus

CEREBELLAR DEEP NUCLEI

Superior
cerebellar
peduncle

Ansiform lobule
Crus I
(HVIIA)

Nodulus
(X)

HEMISPHERE

Dentate
nucleus

Inferior
cerebellar
peduncle

Middle and inferior
cerebellar peduncles

Fourth ventricle
(lateral recess)

Germinal trigone

Ansiform lobule
Crus II
(HVIIA)

Choroid
plexus

Vestibular
nuclear
complex

Middle
cerebellar
peduncle

Biventral
lobule
(HVIII)

Medial
vestibular
nucleus

Fourth
ventricle

Dorsal
longitudinal
fasciculus

MEDULLA

Paramedian
lobule
(HVIIB)

Flocculus
(HX)

Solitary
nucleus and
tract

Dorsal
sensory
nucleus (X)

Part of
section
torn away

Cuneate
nucleus

Dorsal
motor
nucleus (X)

Germinal trigone

Medullary glioepithelium/ependyma

Choroid
plexus

Hypoglossal
nucleus (XII)

Reticular
formation

Spinal
nucleus
and tract
(V)

Nucleus of Roller

Medial
longitudinal
fasciculus and
tectospinal
tract

Stem cells of choroid plexus

External cuneate nucleus

Lateral
reticular
nucleus

Inferior cerebellar peduncle
and spinocerebellar tracts

Inferior
olive complex
(principal
nucleus)

Medial
accessory olive

Pyramid
(corticospinal
tract)

MEDULLA

Spinocephalic tract?

Raphe nuclear complex

Medial lemniscus

Germinal and transitional structures in *italics*

PLATE 55A
CR 260 mm
GW 30
Y14-59
Frontal
Section 1361

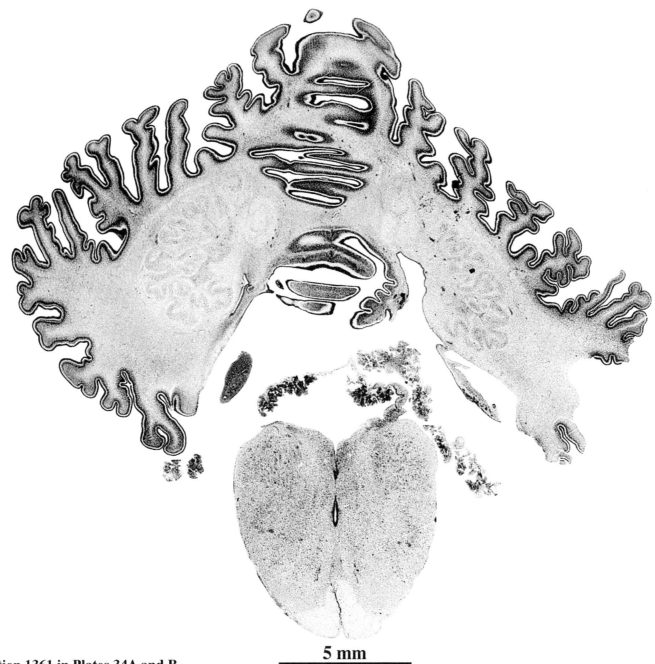

5 mm

See the entire Section 1361 in Plates 34A and B.

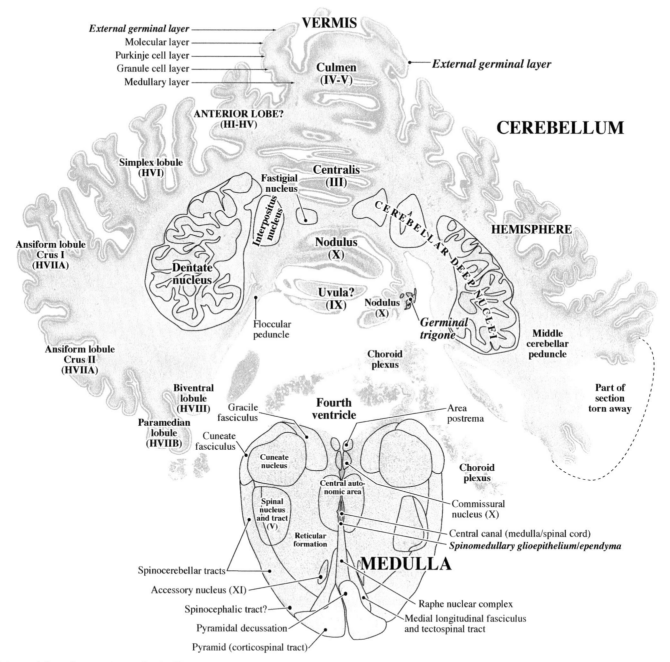

VERMIS

CEREBELLUM

External germinal layer

Molecular layer
Purkinje cell layer
Granule cell layer
Medullary layer

Culmen
(IV-V)

External germinal layer

ANTERIOR LOBE?
(HI-HV)

Simplex lobule
(HVI)

Centralis
(III)

Fastigial
nucleus

*Interpositus
nucleus*

CEREBELLAR-DEEP-NUCLEI

HEMISPHERE

Ansiform lobule
Crus I
(HVIIA)

Nodulus
(X)

Dentate
nucleus

Uvula?
(IX)

Nodulus
(X)

*Germinal
trigone*

Middle
cerebellar
peduncle

Ansiform lobule
Crus II
(HVIIA)

Floccular
peduncle

Choroid
plexus

Part of
section
torn away

Biventral
lobule
(HVIII)

Gracile
fasciculus

Fourth
ventricle

Area
postrema

Paramedian
lobule
(HVIIB)

Cuneate
fasciculus

Cuneate
nucleus

Choroid
plexus

Central auto-
nomic area

Commissural
nucleus (X)

Spinal
nucleus
and tract
(V)

Central canal (medulla/spinal cord)
Spinomedullary glioepithelium/ependyma

Reticular
formation

MEDULLA

Spinocerebellar tracts

Accessory nucleus (XI)

Raphe nuclear complex

Spinocephalic tract?

Medial longitudinal fasciculus
and tectospinal tract

Pyramidal decussation

Pyramid (corticospinal tract)

Germinal and transitional structures in *italics*

PART IV: Y187-65
CR 260 mm (GW 30)
Horizontal

This specimen is case number W-187-65 (Perinatal RPSL) in the Yakovlev Collection. A male infant survived 17 hours after a premature birth. Death occurred because of a pulmonary hyaline membrane. Autopsy notes include a subarachnoid hemorrhage over the cerebrum, but the brain appears normal and is classified as a Normative Control in the Yakovlev Collection (Haleem, 1990). It was cut in the horizontal plane in 35-μm thick sections. Since there is no available photograph of this brain before it was embedded and cut, we use the lateral view of another GW 30 brain that Larroche published in 1967 (**Figure 6**).

The approximate cutting plane of this brain is indicated in **Figure 7** (facing page) with lines superimposed on the GW 30 brain from the Larroche (1967) series. The anterior part of each section is ventral to the posterior part. As in all other specimens, the sections chosen for illustration are more closely spaced to show small structures in the diencephalon, midbrain, pons, and medulla. Illustrated sections are spaced farther apart when they contain only large brain structures, such as the cerebral cortex, basal ganglia, and cerebellum. Low-magnification photographs of Nissl-stained sections that include the cerebral cortex are shown **Plates 56-72**. **Plates 73-79** show high-magnification sections of the brainstem and cerebellum below the cerebral cortex. The core of the brain and the cerebellum are shown at high magnification in **Plates 80-94.**

A densely staining ***neuroepithelium/subventricular zone*** is present and presumably generat-

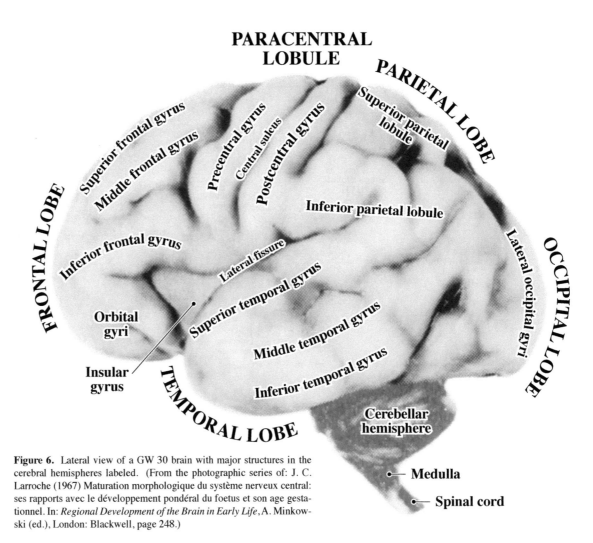

Figure 6. Lateral view of a GW 30 brain with major structures in the cerebral hemispheres labeled. (From the photographic series of: J. C. Larroche (1967) Maturation morphologique du système nerveux central: ses rapports avec le développement pondéral du foetus et son age gestationnel. In: *Regional Development of the Brain in Early Life*, A. Minkowski (ed.), London: Blackwell, page 248.)

ing neocortical interneurons in all lobes of the cerebral cortex. Remnants of migrating and sojourning neurons and/or glia are visible in dwindling ***stratified transitional fields*** of the cerebral cortex. The ***rostral migratory stream*** contains neurons, glia, and their mitotic precursor cells moving through the olfactory peduncle toward the olfactory bulb from a presumed source

area in the germinal matrix at the junction between the cerebral cortex, striatum, and nucleus accumbens. The ***lateral migratory stream*** percolates through the claustrum, endopiriform nucleus, external capsule, and uncinate fasciculus with dense streams of cells that appear to be heading toward the insular cortex, primary olfactory cortex, temporal cortex, and basolateral parts of the

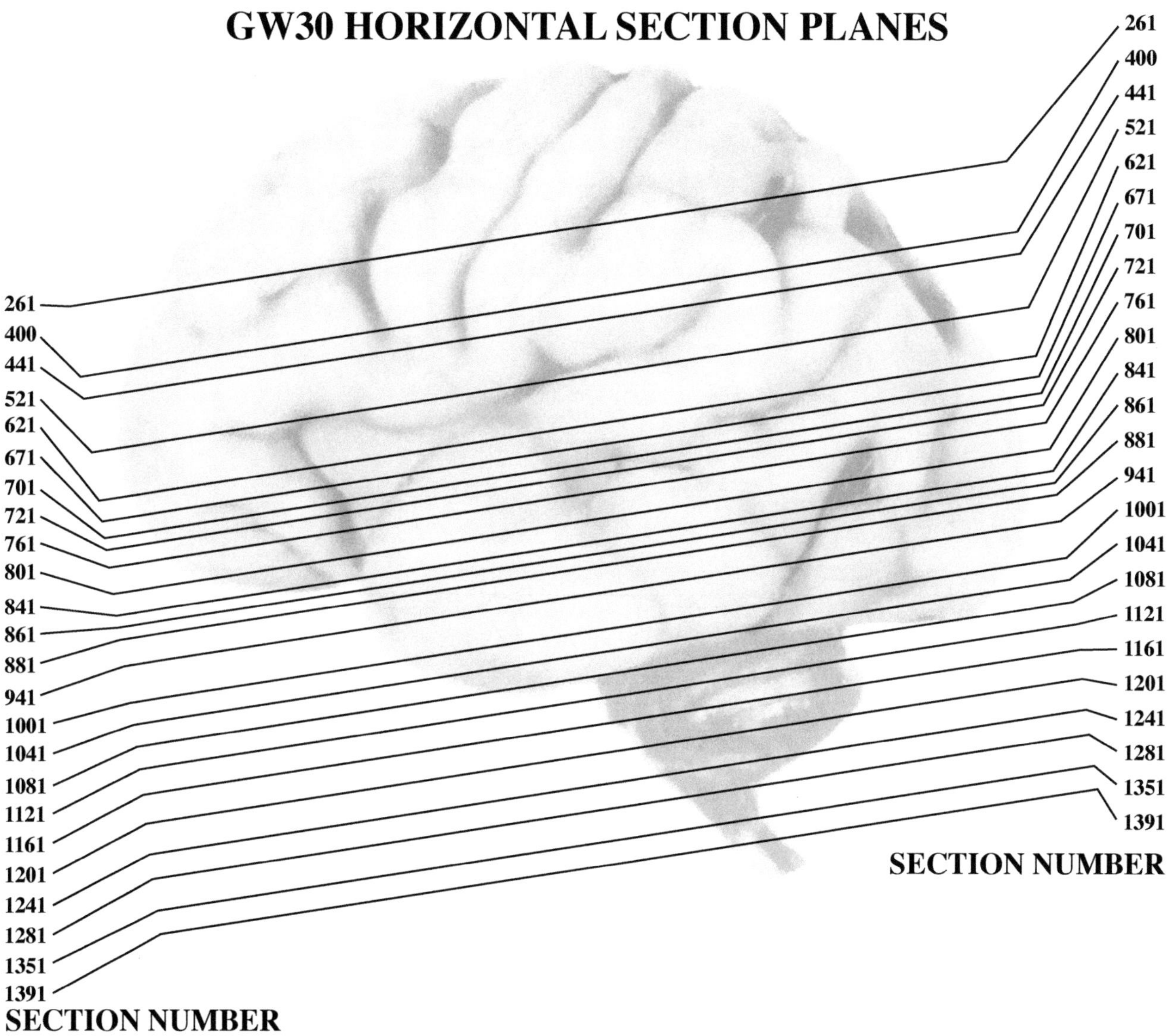

Figure 7. Lateral view of the same GW 30 brain shown in **Figure 6** with the approximate locations and cutting angle of the sections of Y187-65. (From the photographic series of: J. C. Larroche (1967) Maturation morphologique du système nerveux central: ses rapports avec le développement pondéral du foetus et son age gestationnel. In: *Regional Development of the Brain in Early Life*, A. Minkowski (ed.), London: Blackwell, page 248.

amygdaloid complex. In the basal ganglia, there is a large ***neuroepithelium/ subventricular zone*** producing neurons in the striatum and nucleus accumben. At least three subdivisions (anteromedial, anterolateral, and posterior) can be distinguished in the striatal part. Another region of active neurogenesis in the telencephalon is the ***subgranular zone*** in the hilus of the dentate gyrus that is the source of granule cells. Other structures in the telencephalon, such as the septum, fornix, and Ammon's horn, have only a thin, darkly staining layer at the ventricle, and these are presumed to be generating glia, cells of the choroid plexus, and the ependymal lining of the ventricle.

Most of the structures in the diencephalon appear to be settled and are maturing. The third ventricle is lined by a ***glioepithelium/ependyma***. A convoluted ***glioepithelium/ependyma*** lines the cerebral aqueduct in the midbrain that continues into the anterior fourth ventricle. A smooth ***glioepithelium/ependyma*** lines the fourth ventricle through the posterior pons. A slightly convoluted ***glioepithelium/ependyma*** lines the floor of the fourth ventricle through much of the medulla.

The ***external germinal layer*** is prominent over the entire surface of the cerebellar cortex and is actively producing basket, stellate, and granule cells. The ***germinal trigone*** is at the base of the nodulus and along the floccular peduncle; choroid plexus cells and glia are still originating here.

114

PLATE 56A
CR 260 mm
GW 30, Y187-65
Horizontal
Section 261

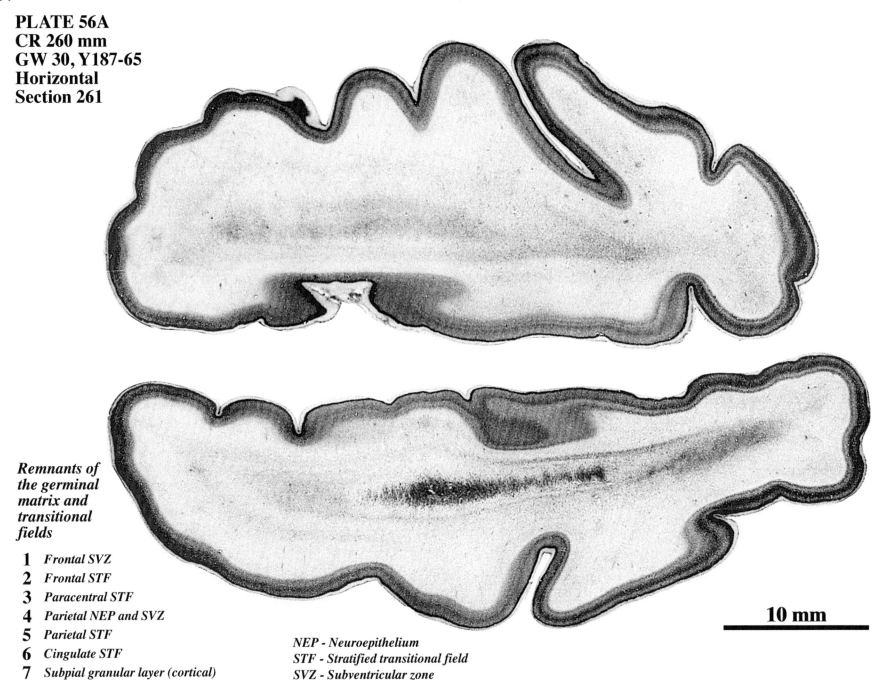

Remnants of
the germinal
matrix and
transitional
fields

1 *Frontal SVZ*
2 *Frontal STF*
3 *Paracentral STF*
4 *Parietal NEP and SVZ*
5 *Parietal STF*
6 *Cingulate STF*
7 *Subpial granular layer (cortical)*

NEP - Neuroepithelium
STF - Stratified transitional field
SVZ - Subventricular zone

10 mm

PLATE 56B

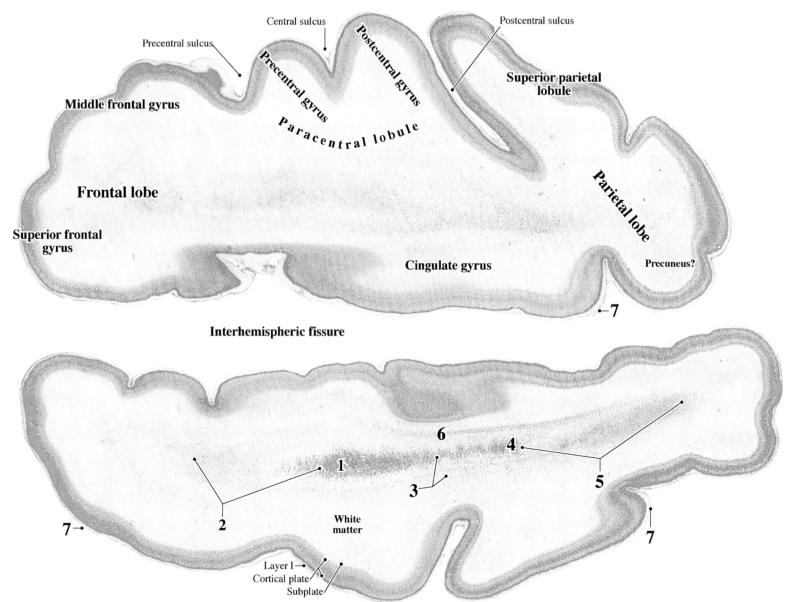

Central sulcus

Precentral sulcus

Postcentral sulcus

Precentral gyrus

Postcentral gyrus

Superior parietal lobule

Middle frontal gyrus

Paracentral lobule

Frontal lobe

Parietal lobe

Superior frontal gyrus

Cingulate gyrus

Precuneus?

7

Interhemispheric fissure

6

4

1

3

5

2

7

White matter

Layer I

Cortical plate

Subplate

7

PLATE 57A
CR 260 mm
GW 30, Y187-65
Horizontal
Section 400

Remnants of
the germinal
matrix and
transitional
fields

1 *Frontal NEP and SVZ*
2 *Frontal STF*
3 *Callosal GEP*
4 *Callosal sling*
5 *Fornical GEP*
6 *Parietal NEP and SVZ*
7 *Parietal STF*
8 *Posterior striatal NEP and SVZ*
9 *Anterolateral striatal NEP and SVZ*
10 *Anteromedial striatal NEP and SVZ*
11 *Strionuclear GEP*
12 *Subpial granular layer (cortical)*

GEP - Glioepithelium
NEP - Neuroepithelium
STF - Stratified transitional field
SVZ - Subventricular zone

10 mm

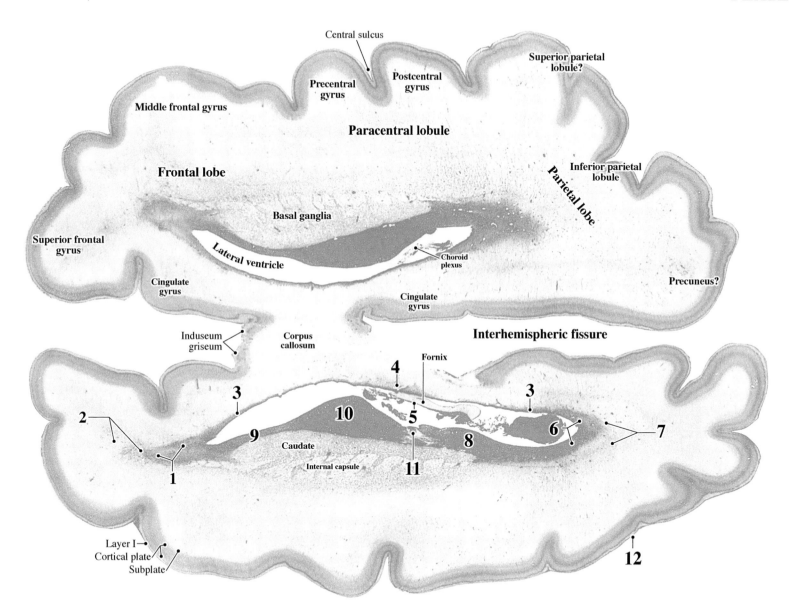

Central sulcus

Superior parietal
lobule?

Precentral
gyrus

Postcentral
gyrus

Middle frontal gyrus

Paracentral lobule

Frontal lobe

Inferior parietal
lobule

Parietal lobe

Basal ganglia

*Superior frontal
gyrus*

Lateral ventricle

Choroid
plexus

Precuneus?

Cingulate
gyrus

Cingulate
gyrus

Interhemispheric fissure

Induseum
griseum

Corpus
callosum

Fornix

4

3

3

2

10

5

6

7

9

Caudate

8

Internal capsule

11

1

Layer I

Cortical plate

Subplate

12

PLATE 58A
CR 260 mm
GW 30, Y187-65
Horizontal
Section 441

See detail of the brain core
in Plates 80A and B.

*Remnants of
the germinal
matrix and
transitional
fields*

1 *Frontal NEP and SVZ*
2 *Frontal STF*
3 *Callosal GEP*
4 *Callosal sling*
5 *Fornical GEP*
6 *Parietal NEP and SVZ*
7 *Parietal STF*
8 *Posterior striatal NEP and SVZ*
9 *Anterolateral striatal NEP and SVZ*
10 *Anteromedial striatal NEP and SVZ*
11 *Strionuclear GEP*
12 *Subpial granular layer (cortical)*

GEP - Glioepithelium
NEP - Neuroepithelium
STF - Stratified transitional field
SVZ - Subventricular zone

10 mm

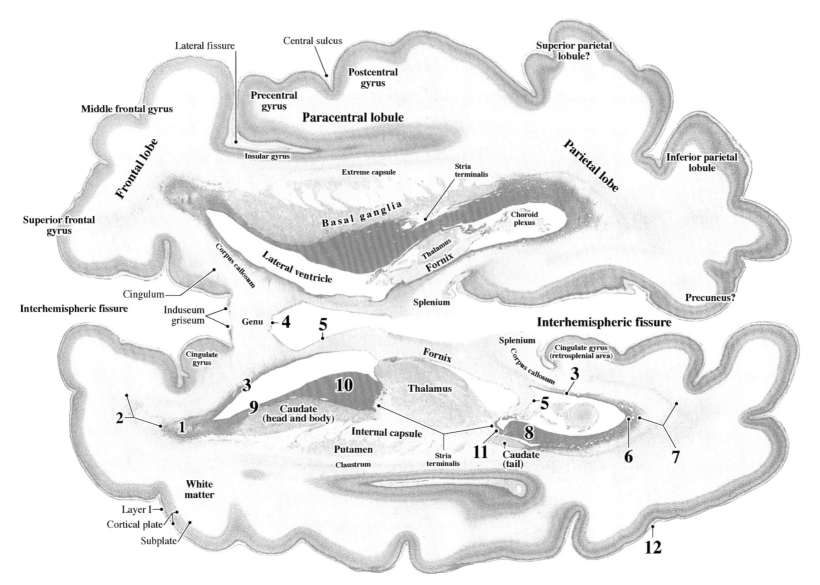

Lateral fissure

Central sulcus

Postcentral gyrus

Superior parietal lobule?

Precentral gyrus

Middle frontal gyrus

Paracentral lobule

Insular gyrus

Frontal lobe

Extreme capsule

Stria terminalis

Parietal lobe

Inferior parietal lobule

Superior frontal gyrus

Basal ganglia

Choroid plexus

Lateral ventricle

Thalamus

Fornix

Corpus callosum

Cingulum

Splenium

Interhemispheric fissure

Induseum griseum

Genu **4**

5

Interhemispheric fissure

Precuneus?

Cingulate gyrus

Fornix

Splenium

Cingulate gyrus (retrosplenial area)

Corpus callosum

3

3

9

10

Thalamus

5

2

Caudate (head and body)

Internal capsule

8

1

Putamen

Stria terminalis

11

Caudate (tail)

6

7

Claustrum

White matter

Layer I

Cortical plate

Subplate

12

PLATE 59A
CR 260 mm
GW 30, Y187-65
Horizontal
Section 521

See detail of the brain core
in Plates 81A and B.

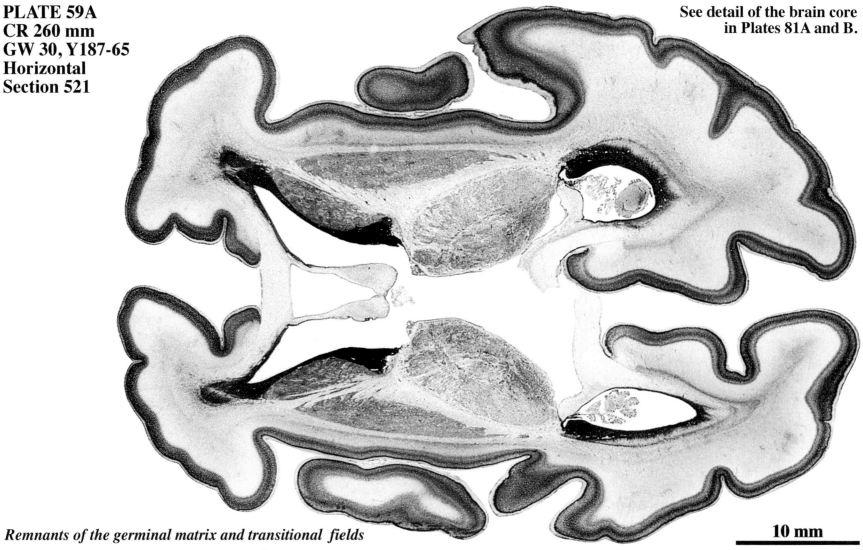

10 mm

Remnants of the germinal matrix and transitional fields

1 *Rostral migratory stream (source area)*
2 *Frontal NEP and SVZ*
3 *Frontal STF*
4 *Callosal GEP*
5 *Callosal sling*
6 *Fornical GEP*
7 *Parietal NEP and SVZ*

8 *Parietal STF*
9 *Posterior striatal NEP and SVZ*
10 *Anterolateral striatal NEP and SVZ*
11 *Anteromedial striatal NEP and SVZ*
12 *Strionuclear GEP*
13 *Septal G/EP*
14 *Subpial granular layer (cortical)*

GEP - Glioepithelium
G/EP - Glioepithelium/ependyma
NEP - Neuroepithelium
STF - Stratified transitional field
SVZ - Subventricular zone

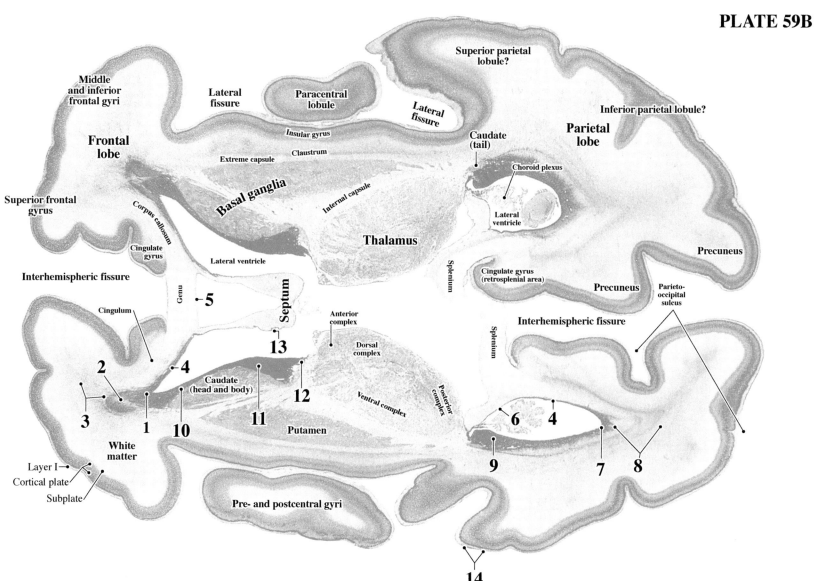

Middle
and inferior
frontal gyri

Lateral
fissure

Paracentral
lobule

Lateral
fissure

Superior parietal
lobule?

Inferior parietal lobule?

Frontal
lobe

Insular gyrus

Claustrum

Extreme capsule

Caudate
(tail)

Choroid plexus

Parietal
lobe

Superior frontal
gyrus

Basal ganglia

Internal capsule

Lateral
ventricle

Corpus callosum

Thalamus

Cingulate
gyrus

Splenium

Precuneus

Lateral ventricle

Interhemispheric fissure

Cingulate gyrus
(retrosplenial area)

Precuneus

Parieto-
occipital
sulcus

Genu

Septum

Splenium

Interhemispheric fissure

5

Cingulum

Anterior
complex

13

Dorsal
complex

2

4

12

Caudate
(head and body)

Ventral complex

Posterior
complex

3

1

11

6

4

10

Putamen

9

7

8

White
matter

Layer I →

Cortical plate

Subplate

Pre- and postcentral gyri

14

122

See detail of the brain core
in Plates 82A and B.

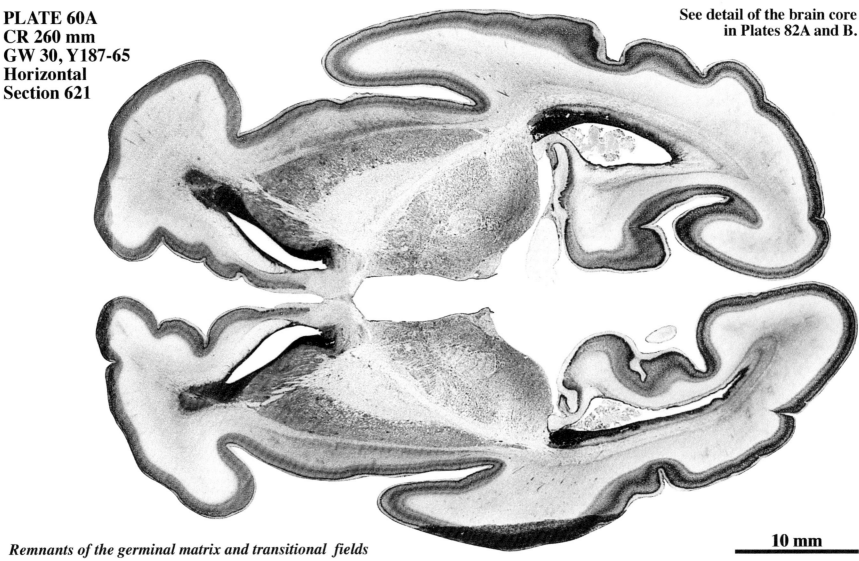

10 mm

Remnants of the germinal matrix and transitional fields

1 *Rostral migratory stream (source area)*

2 *Frontal NEP and SVZ*

3 *Frontal STF*

4 *Callosal GEP*

5 *Fornical GEP*

6 *Parahippocampal NEP, SVZ, and STF*

7 *Occipital NEP and SVZ*

8 *Occipital STF*

9 *Parietal/temporal NEP and SVZ*

10 *Parietal/temporal STF*

11 *Alvear GEP*

12 *Subgranular zone (dentate)*

13 *Posterior striatal NEP and SVZ*

14 *Anterolateral striatal NEP and SVZ*

15 *Anteromedial striatal NEP and SVZ*

16 *Strionuclear NEP and SVZ*

17 *Septal G/EP*

18 *Subpial granular layer (cortical)*

GEP - Glioepithelium
G/EP - Glioepithelium/ependyma
NEP - Neuroepithelium
STF - Stratified transitional field
SVZ - Subventricular zone

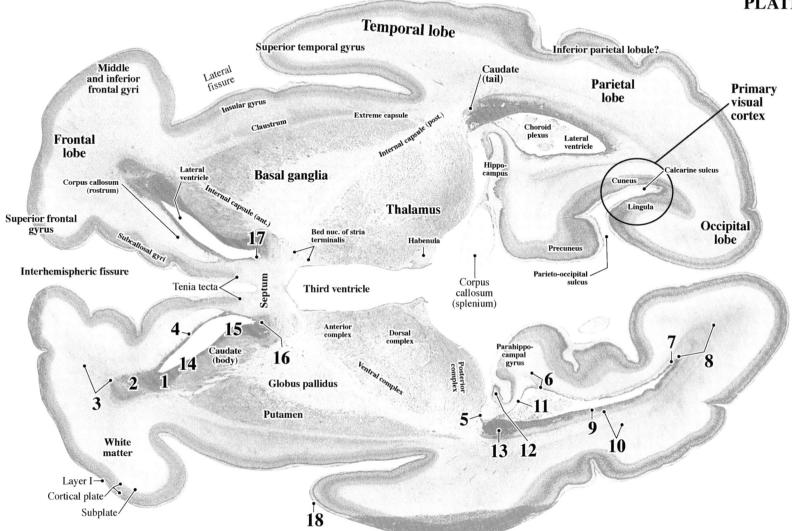

PLATE 61A
CR 260 mm
GW 30, Y187-65
Horizontal
Section 671

See detail of the brain core
in Plates 83A and B.

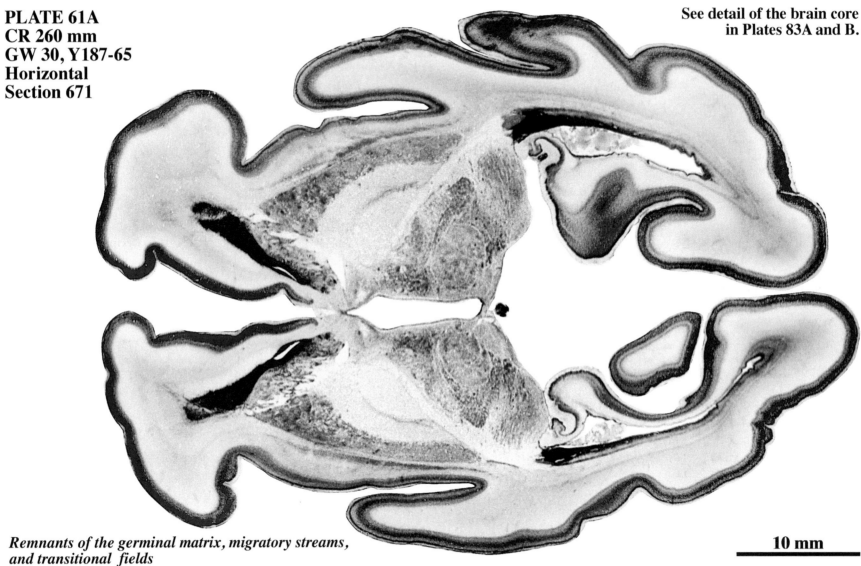

Remnants of the germinal matrix, migratory streams, and transitional fields

1 Rostral migratory stream (source area)
2 Frontal NEP and SVZ
3 Frontal STF
4 Callosal GEP
5 Fornical GEP
6 Parahippocampal NEP, SVZ, and STF

7 Occipital NEP and SVZ
8 Occipital STF
9 Temporal NEP and SVZ
10 Temporal STF
11 Alvear GEP
12 Subgranular zone (dentate)

13 Lateral migratory stream (cortical)
14 Posterior striatal NEP and SVZ
15 Anterolateral striatal NEP and SVZ
   (infiltrated by the rostral migratory stream)
16 Accumbent NEP and SVZ
   (infiltrated by the rostral migratory stream)
17 Subpial granular layer (cortical)

GEP - Glioepithelium
G/EP - Glioepithelium/ependyma
NEP - Neuroepithelium
STF - Stratified transitional field
SVZ - Subventricular zone

10 mm

125

PLATE 61B

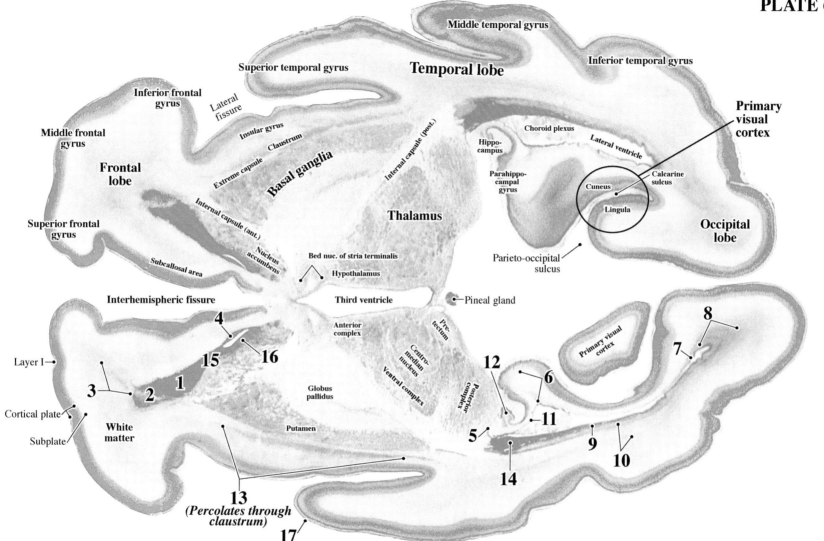

Middle temporal gyrus

Superior temporal gyrus

Temporal lobe

Inferior temporal gyrus

Inferior frontal gyrus

Lateral fissure

Insular gyrus

Claustrum

Extreme capsule

Basal ganglia

Internal capsule (post.)

Middle frontal gyrus

Frontal lobe

Superior frontal gyrus

Internal capsule (ant.)

Thalamus

Choroid plexus

Hippo-campus

Lateral ventricle

Parahippo-campal gyrus

Cuneus

Calcarine sulcus

Lingula

Primary visual cortex

Occipital lobe

Subcallosal area

Nucleus accumbens

Bed nuc. of stria terminalis

Hypothalamus

Parieto-occipital sulcus

Interhemispheric fissure

Third ventricle

Pineal gland

Anterior complex

Pre-tectum

Layer I

Centro-median nucleus

Ventral complex

Posterior complex

Primary visual cortex

Cortical plate

Globus pallidus

Subplate

White matter

Putamen

13 (Percolates through claustrum)

17

PLATE 62A
CR 260 mm
GW 30, Y187-65
Horizontal
Section 701

See detail of the brain core
in Plates 84A and B.

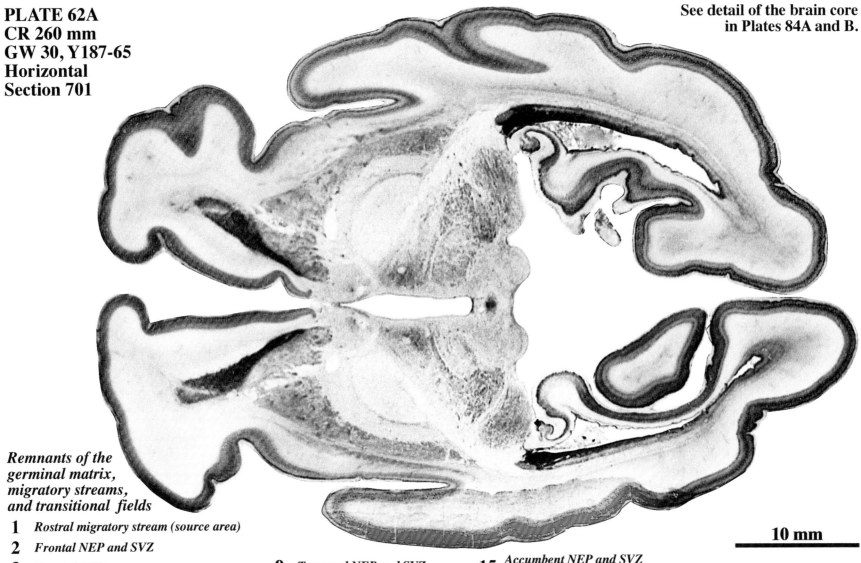

Remnants of the
germinal matrix,
migratory streams,
and transitional fields

1   *Rostral migratory stream (source area)*
2   *Frontal NEP and SVZ*
3   *Frontal STF*
4   *Callosal GEP*
5   *Fornical GEP*
6   *Parahippocampal NEP, SVZ, and STF*
7   *Occipital NEP and SVZ*
8   *Occipital STF*

9   *Temporal NEP and SVZ*
10 *Temporal STF*
11 *Alvear GEP*
12 *Subgranular zone (dentate)*
13 *Lateral migratory stream (cortical)*
14 *Posterior striatal NEP and SVZ*

15 *Accumbent NEP and SVZ*
     (infiltrated by the rostral migratory stream)
16 *Subpial granular layer (cortical)*

GEP - Glioepithelium
NEP - Neuroepithelium
STF - Stratified transitional field
SVZ - Subventricular zone

10 mm

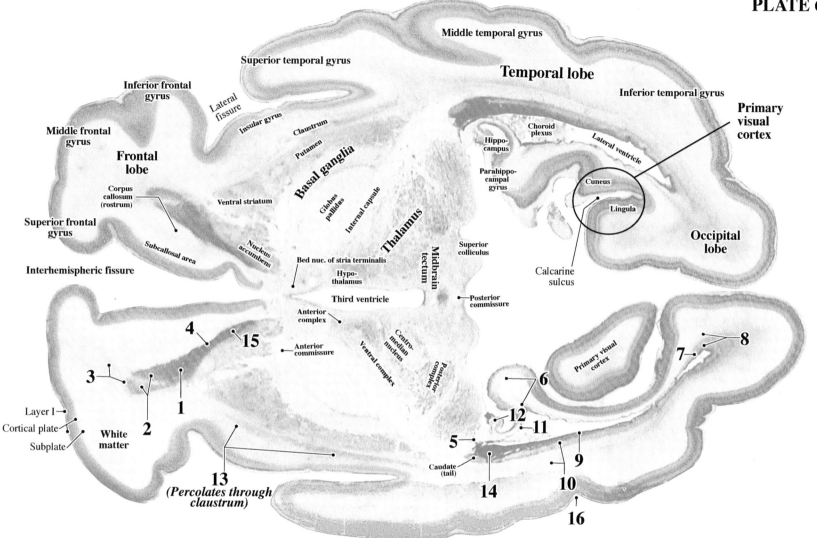

Middle temporal gyrus

Superior temporal gyrus

Temporal lobe

Inferior frontal gyrus

Lateral fissure

Insular gyrus

Inferior temporal gyrus

Primary visual cortex

Middle frontal gyrus

Claustrum

Putamen

Choroid plexus

Hippo-campus

Lateral ventricle

Frontal lobe

Basal ganglia

Corpus callosum (rostrum)

Parahippo-campal gyrus

Cuneus

Superior frontal gyrus

Ventral striatum

Globus pallidus

Internal capsule

Thalamus

Lingula

Occipital lobe

Subcallosal area

Nucleus accumbens

Bed nuc. of stria terminalis

Superior colliculus

Calcarine sulcus

Interhemispheric fissure

Hypo-thalamus

Midbrain tectum

Third ventricle

Posterior commissure

Anterior complex

4

15

Anterior commissure

Centro-median nucleus

Ventral complex

Posterior complex

8

7

3

Primary visual cortex

1

6

Layer I

2

12

11

Cortical plate

5

9

Subplate

White matter

Caudate (tail)

10

13
(Percolates through claustrum)

14

16

PLATE 63A
CR 260 mm
GW 30, Y187-65
Horizontal
Section 721

See detail of the brain core
in Plates 85A and B.

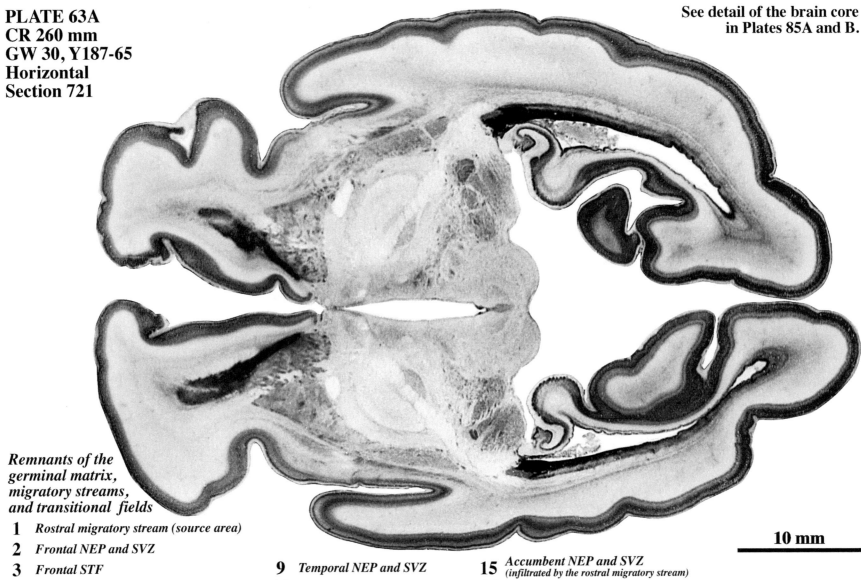

10 mm

*Remnants of the
germinal matrix,
migratory streams,
and transitional fields*

1 *Rostral migratory stream (source area)*

2 *Frontal NEP and SVZ*

3 *Frontal STF*

4 *Callosal GEP*

5 *Fornical GEP*

6 *Parahippocampal NEP, SVZ, and STF*

7 *Occipital NEP and SVZ*

8 *Occipital STF*

9 *Temporal NEP and SVZ*

10 *Temporal STF*

11 *Alvear GEP*

12 *Subgranular zone (dentate)*

13 *Lateral migratory stream (cortical)*

14 *Posterior striatal NEP and SVZ*

15 *Accumbent NEP and SVZ*
(infiltrated by the rostral migratory stream)

16 *Subpial granular layer (cortical)*

GEP - Glioepithelium
NEP - Neuroepithelium
STF - Stratified transitional field
SVZ - Subventricular zone

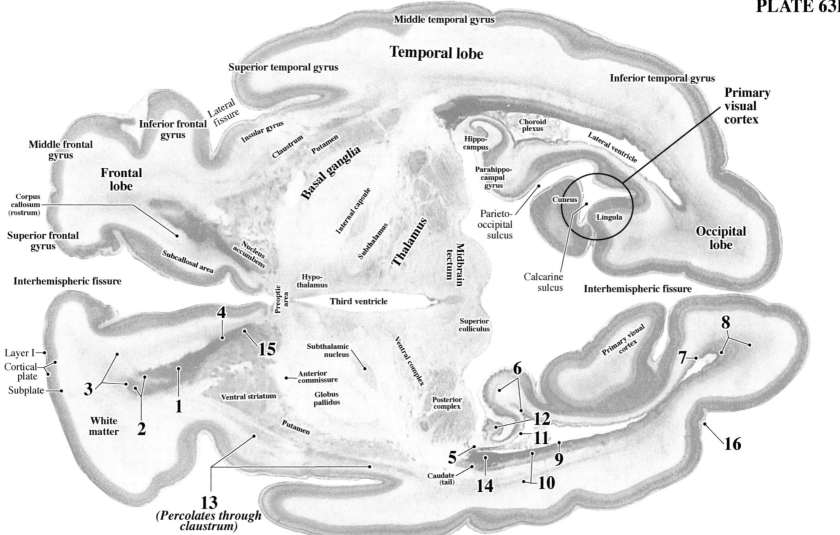

Middle temporal gyrus

Temporal lobe

Superior temporal gyrus

Inferior temporal gyrus

Primary visual cortex

Lateral fissure

Inferior frontal gyrus

Insular gyrus

Choroid plexus

Hippo-campus

Lateral ventricle

Claustrum

Putamen

Middle frontal gyrus

Frontal lobe

Basal ganglia

Parahippo-campal gyrus

Corpus callosum (rostrum)

Internal capsule

Cuneus

Lingula

Superior frontal gyrus

Subcallosal area

Nucleus accumbens

Subthalamus

Thalamus

Parieto-occipital sulcus

Occipital lobe

Hypo-thalamus

Midbrain tectum

Calcarine sulcus

Interhemispheric fissure

Preoptic area

Third ventricle

Superior colliculus

Interhemispheric fissure

4

15

Subthalamic nucleus

Ventral complex

8

Layer I

Cortical plate

Subplate

3

Anterior commissure

Posterior complex

Primary visual cortex

7

6

1

Ventral striatum

Globus pallidus

2

White matter

12

5

11

9

Putamen

16

Caudate (tail)

14

10

13
(Percolates through claustrum)

130

See detail of the brain core
in Plates 86A and B.

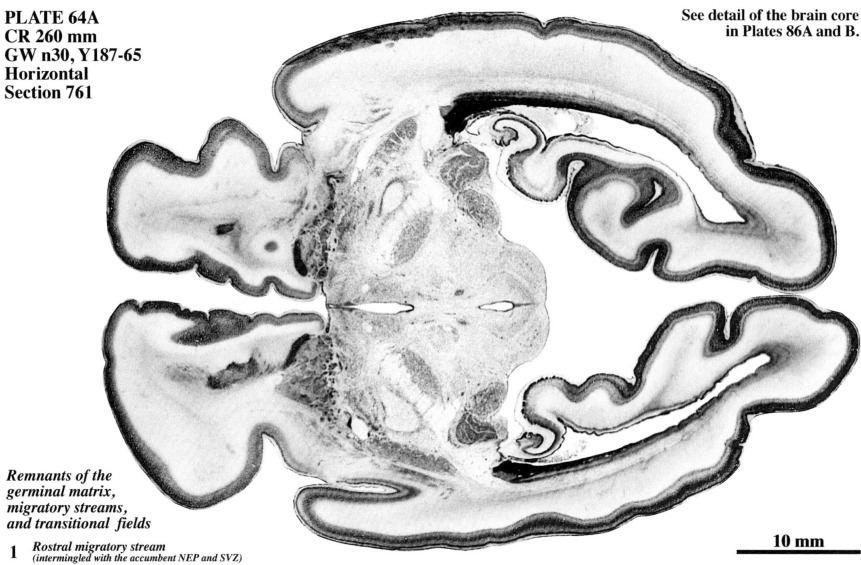

10 mm

*Remnants of the
germinal matrix,
migratory streams,
and transitional fields*

1 **Rostral migratory stream**
   (intermingled with the accumbent NEP and SVZ)
2 **Frontal NEP and SVZ**
3 **Frontal STF**
4 **Fornical GEP**
5 **Parahippocampal NEP, SVZ, and STF**
6 **Occipital NEP and SVZ**
7 **Occipital STF**

8 **Temporal NEP and SVZ**
9 **Temporal STF**
10 **Alvear GEP**
11 **Subgranular zone (dentate)**
12 **Lateral migratory stream (cortical)**

13 **Posterior striatal NEP and SVZ**
14 **Subpial granular layer (cortical)**

GEP - Glioepithelium
NEP - Neuroepithelium
STF - Stratified transitional field
SVZ - Subventricular zone

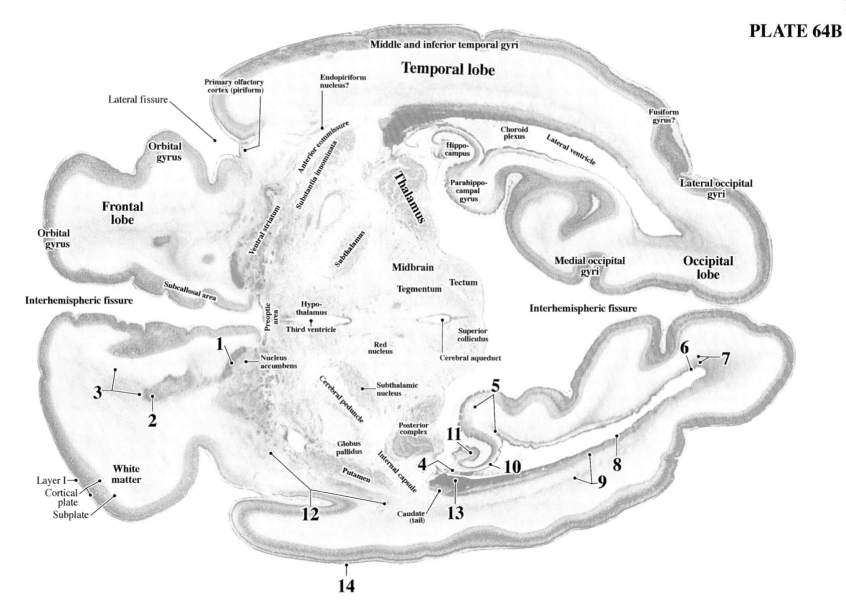

Middle and inferior temporal gyri

Temporal lobe

Primary olfactory
cortex (piriform)

Endopiriform
nucleus?

Lateral fissure

Choroid
plexus

Fusiform
gyrus?

**Orbital
gyrus**

Hippo-
campus

Lateral ventricle

Anterior commissure

**Frontal
lobe**

Substantia innominata

Parahippo-
campal
gyrus

Thalamus

**Lateral occipital
gyri**

Orbital
gyrus

Ventral striatum

Subthalamus

Midbrain

**Medial occipital
gyri**

**Occipital
lobe**

Subcallosal area

Tectum

Interhemispheric fissure

Preoptic
area

Hypo-
thalamus

Tegmentum

Interhemispheric fissure

Third ventricle

Superior
colliculus

1

Nucleus
accumbens

Red
nucleus

Cerebral aqueduct

6

7

3

Cerebral peduncle

Subthalamic
nucleus

5

2

11

Posterior
complex

10

4

8

Globus
pallidus

Internal capsule

13

9

**White
matter**

Putamen

Layer I
Cortical
plate
Subplate

12

Caudate
(tail)

14

132

PLATE 65A
CR 260 mm
GW 30, Y187-65
Horizontal
Section 801

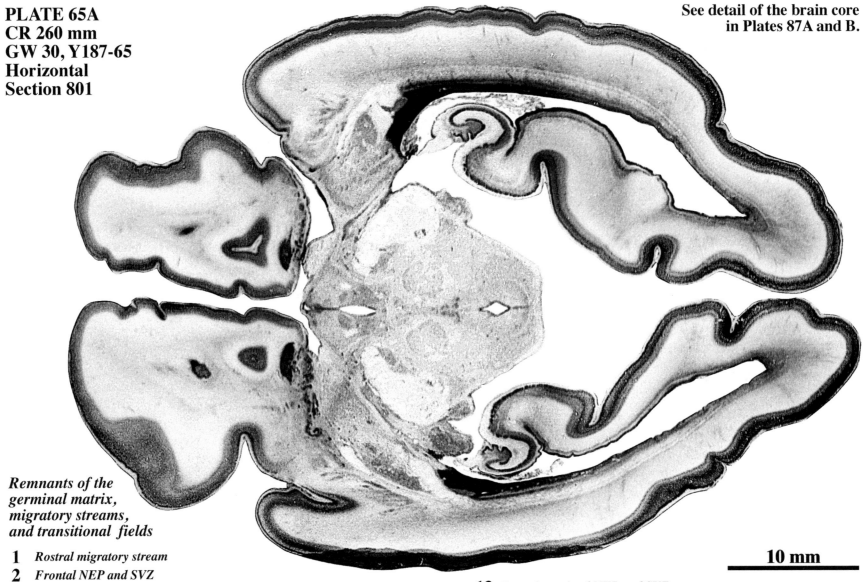

Remnants of the
germinal matrix,
migratory streams,
and transitional fields

1 *Rostral migratory stream*
2 *Frontal NEP and SVZ*
3 *Frontal STF*
4 *Fornical GEP*
5 *Parahippocampal NEP, SVZ, and STF*
6 *Occipital NEP and SVZ*
7 *Occipital STF*

8 *Temporal NEP and SVZ*
9 *Temporal STF*
10 *Alvear GEP*
11 *Subgranular zone (dentate)*
12 *Lateral migratory stream (cortical)*

13 *Posterior striatal NEP and SVZ*
14 *Subpial granular layer (cortical)*

10 mm

GEP - Glioepithelium
NEP - Neuroepithelium
STF - Stratified transitional field
SVZ - Subventricular zone

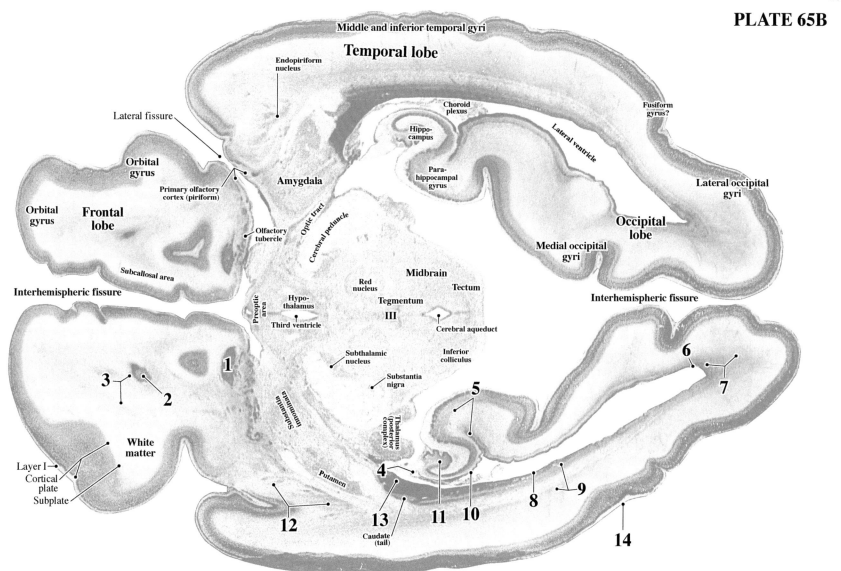

Middle and inferior temporal gyri

Temporal lobe

Endopiriform nucleus

Choroid plexus

Hippo-campus

Lateral ventricle

Fusiform gyrus?

Lateral fissure

Para-hippocampal gyrus

Orbital gyrus

Amygdala

Primary olfactory cortex (piriform)

Optic tract

Cerebral peduncle

Olfactory tubercle

Lateral occipital gyri

Orbital gyrus

Frontal lobe

Red nucleus

Midbrain

Tectum

Occipital lobe

Subcallosal area

Preoptic area

Hypo-thalamus

Tegmentum

III

Medial occipital gyri

Interhemispheric fissure

Third ventricle

Cerebral aqueduct

Interhemispheric fissure

1

Subthalamic nucleus

Inferior colliculus

6

3

Substantia nigra

7

2

Substantia innominata

5

White matter

Thalamus (posterior complex)

Layer I
Cortical plate
Subplate

4

9

Putamen

8

12

13

11

10

Caudate (tail)

14

PLATE 66A
CR 260 mm
GW 30, Y187-65
Horizontal
Section 841

See detail of the brain core
in Plates 88A and B.

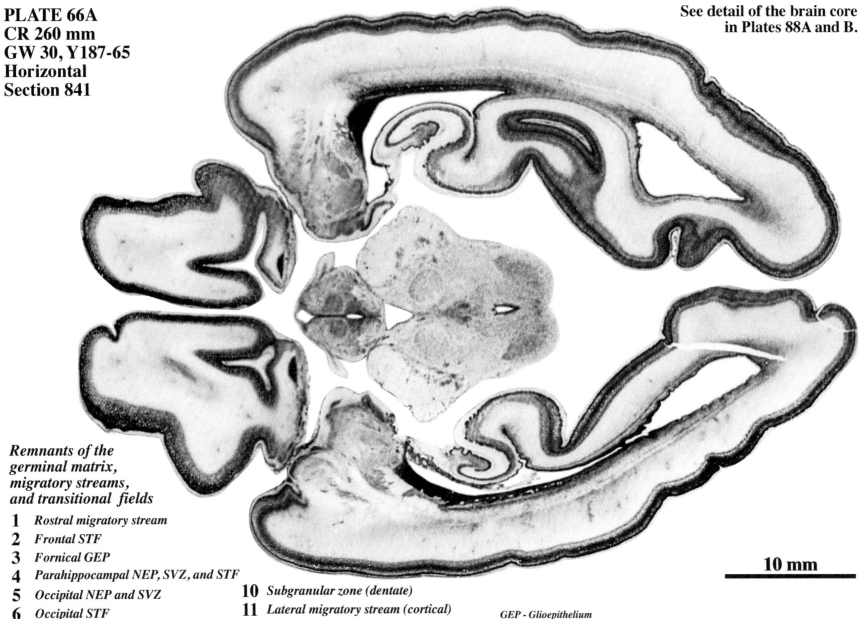

Remnants of the
germinal matrix,
migratory streams,
and transitional fields

1 *Rostral migratory stream*
2 *Frontal STF*
3 *Fornical GEP*
4 *Parahippocampal NEP, SVZ, and STF*
5 *Occipital NEP and SVZ*
6 *Occipital STF*
7 *Temporal NEP and SVZ*
8 *Temporal STF*
9 *Alvear GEP*

10 *Subgranular zone (dentate)*
11 *Lateral migratory stream (cortical)*
12 *Posterior striatal NEP and SVZ*
13 *Amygdaloid G/EP*
14 *Subpial granular layer (cortical)*

GEP - Glioepithelium
G/EP - Glioepithelium/ependyma
NEP - Neuroepithelium
STF - Stratified transitional field
SVZ - Subventricular zone

10 mm

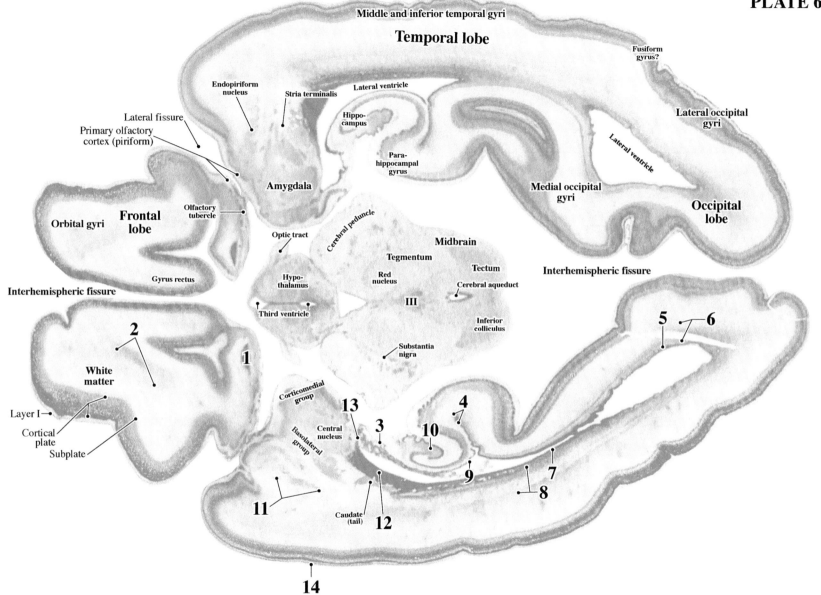

Middle and inferior temporal gyri

Temporal lobe

Fusiform
gyrus?

Endopiriform
nucleus

Stria terminalis

Lateral ventricle

**Lateral occipital
gyri**

Hippo-
campus

Lateral fissure

Primary olfactory
cortex (piriform)

Lateral ventricle

Para-
hippocampal
gyrus

Amygdala

**Medial occipital
gyri**

**Occipital
lobe**

Olfactory
tubercle

**Frontal
lobe**

Orbital gyri

Optic tract

Cerebral peduncle

Midbrain

Gyrus rectus

Hypo-
thalamus

Tegmentum

Red
nucleus

Tectum

Cerebral aqueduct

Interhemispheric fissure

Interhemispheric fissure

Third ventricle

III

Inferior
colliculus

5

6

Substantia
nigra

2

1

**White
matter**

4

Layer I

Corticomedial
group

13

3

10

Cortical
plate

Central
nucleus

9

7

Subplate

Basolateral
group

8

11

Caudate
(tail)

12

14

PLATE 67A
CR 260 mm
GW 30, Y187-65
Horizontal
Section 861

See detail of the brain core
in Plates 89A and B.

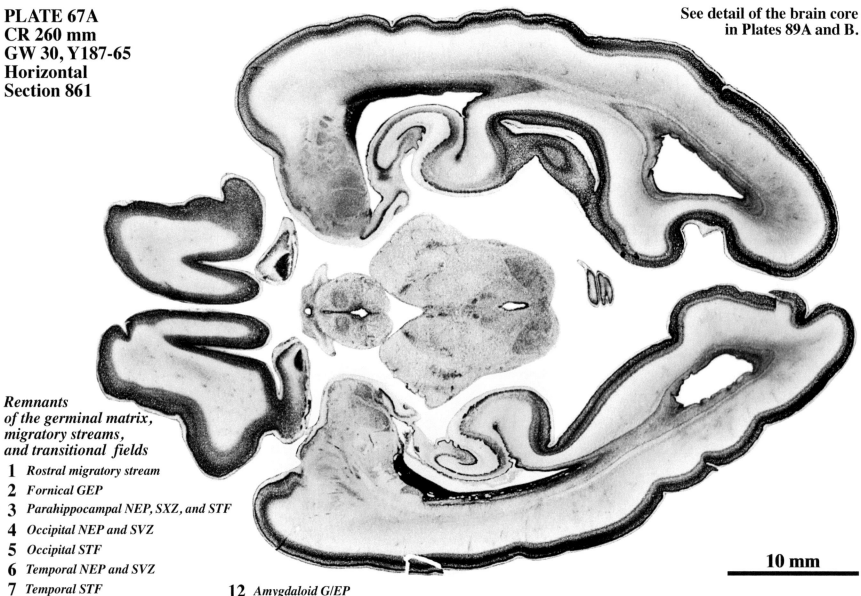

Remnants
of the germinal matrix,
migratory streams,
and transitional fields

1 *Rostral migratory stream*

2 *Fornical GEP*

3 *Parahippocampal NEP, SXZ, and STF*

4 *Occipital NEP and SVZ*

5 *Occipital STF*

6 *Temporal NEP and SVZ*

7 *Temporal STF*

8 *Alvear GEP*

9 *Subgranular zone (dentate)*

10 *Lateral migratory stream (cortical)*

11 *Posterior striatal NEP and SVZ*

12 *Amygdaloid G/EP*

13 *Hypothalamic G/EP*

14 *Mesencephalic G/EP*

15 *External germinal layer (cerebellum)*

16 *Subpial granular layer (cortical)*

GEP - Glioepithelium
G/EP - Glioepithelium/ependyma
NEP - Neuroepithelium
STF - Stratified transitional field
SVZ - Subventricular zone

10 mm

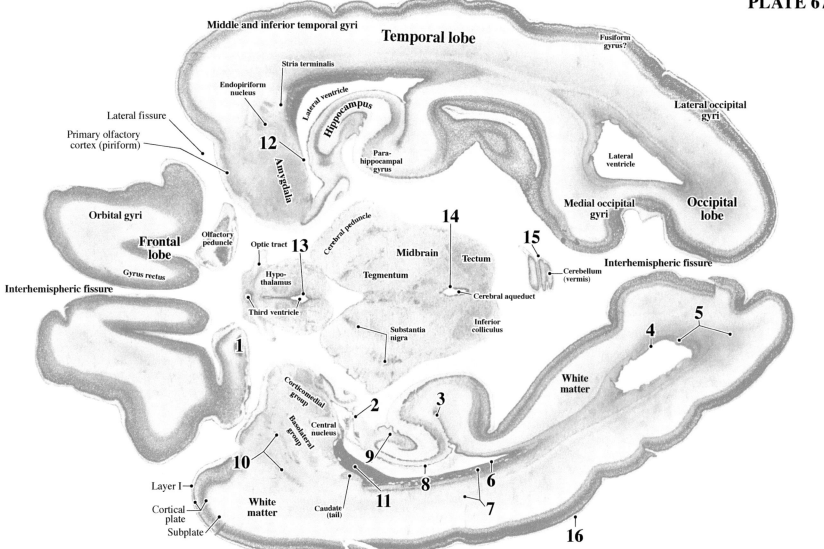

Middle and inferior temporal gyri

Temporal lobe

Fusiform gyrus?

Stria terminalis

Endopiriform nucleus

Lateral ventricle

Hippocampus

Lateral occipital gyri

Lateral fissure

Primary olfactory cortex (piriform)

12

Amygdala

Para-hippocampal gyrus

Lateral ventricle

Medial occipital gyri

Occipital lobe

Orbital gyri

Olfactory peduncle

Optic tract

13

Cerebral peduncle

14

Midbrain

Tectum

15

Cerebellum (vermis)

Frontal lobe

Hypo-thalamus

Tegmentum

Interhemispheric fissure

Gyrus rectus

Third ventricle

Cerebral aqueduct

Interhemispheric fissure

Substantia nigra

Inferior colliculus

1

5

4

White matter

Corticomedial group

2

3

Central nucleus

Basolateral group

9

10

8

6

Layer I

11

7

Cortical plate

White matter

Caudate (tail)

Subplate

16

138

See detail of the brain core
in Plates 90A and B.

PLATE 68A
CR 260 mm
GW 30, Y187-65
Horizontal
Section 881

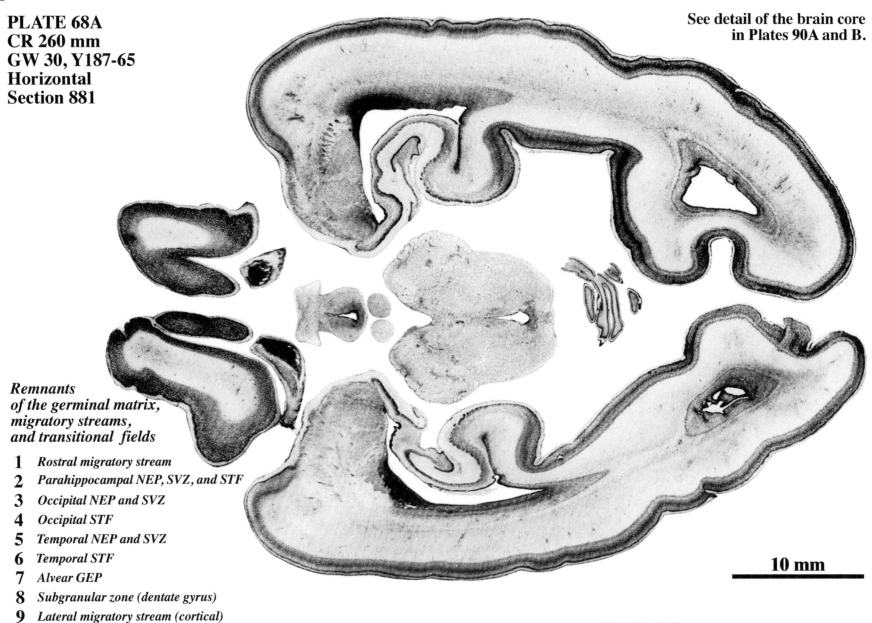

Remnants
of the germinal matrix,
migratory streams,
and transitional fields

1 *Rostral migratory stream*
2 *Parahippocampal NEP, SVZ, and STF*
3 *Occipital NEP and SVZ*
4 *Occipital STF*
5 *Temporal NEP and SVZ*
6 *Temporal STF*
7 *Alvear GEP*
8 *Subgranular zone (dentate gyrus)*
9 *Lateral migratory stream (cortical)*
10 *Amygdaloid G/EP*
11 *Posterior striatal NEP and SVZ*
12 *Hypothalamic G/EP*
13 *Mesencephalic G/EP*
14 *External germinal layer (cerebellum)*
15 *Subpial granular layer (cortical)*

GEP - Glioepithelium
G/EP - Glioepithelium/ependyma
NEP - Neuroepithelium
STF - Stratified transitional field
SVZ - Subventricular zone

10 mm

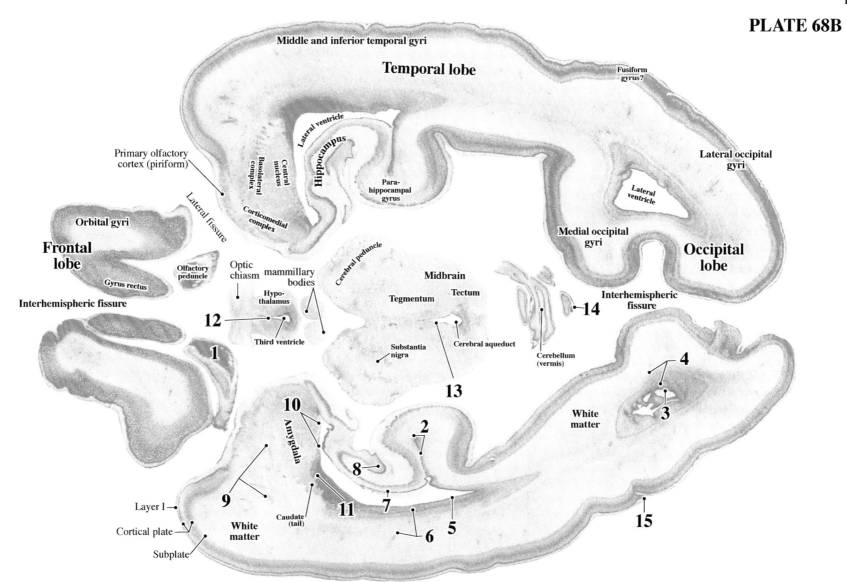

Middle and inferior temporal gyri

Temporal lobe

Fusiform gyrus?

Lateral ventricle

Hippocampus

Lateral occipital gyri

Primary olfactory cortex (piriform)

Basolateral complex

Central nucleus

Para-hippocampal gyrus

Lateral ventricle

Lateral fissure

Corticomedial complex

Orbital gyri

Medial occipital gyri

Frontal lobe

Olfactory peduncle

Optic chiasm

mammillary bodies

Cerebral peduncle

Midbrain

Occipital lobe

Gyrus rectus

Hypo-thalamus

Tegmentum

Tectum

Interhemispheric fissure

Interhemispheric fissure

12

Third ventricle

Substantia nigra

Cerebral aqueduct

Cerebellum (vermis)

14

1

13

White matter

4

10

3

2

Amygdala

White matter

8

9

7

Layer I

Caudate (tail)

11

5

15

Cortical plate

White matter

6

Subplate

PLATE 69A
CR 260 mm
GW 30, Y187-65
Horizontal
Section 941

See detail of the brain core
in Plates 91A and B.

Remnants
of the germinal matrix,
migratory streams,
and transitional fields

1 *Parahippocampal NEP, SVZ, and STF*

2 *Occipital STF*

3 *Temporal NEP and SVZ*

4 *Temporal STF*

5 *Alvear GEP*

6 *Subgranular zone (dentate)*

7 *Lateral migratory stream (cortical)*

8 *Amygdaloid G/EP*

9 *Pontine G/EP*

10 *External germinal layer (cerebellum)*

11 *Subpial granular layer (cortical)*

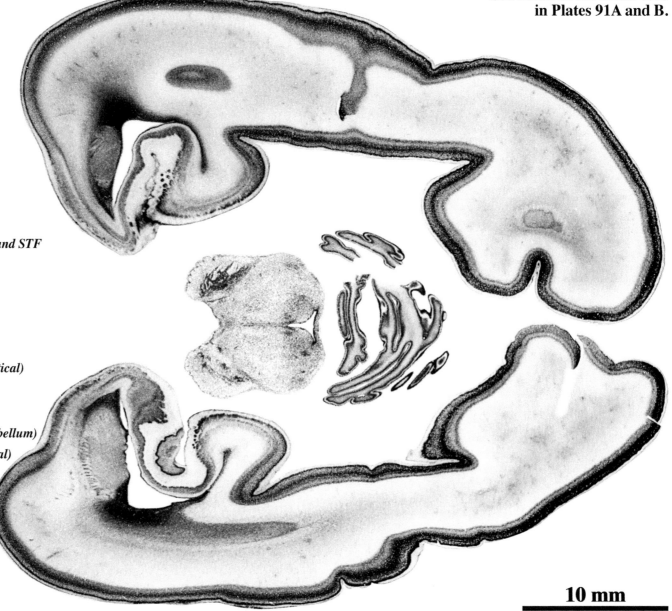

GEP - Glioepithelium
G/EP - Glioepithelium/ependyma
NEP - Neuroepithelium
STF - Stratified transitional field
SVZ - Subventricular zone

10 mm

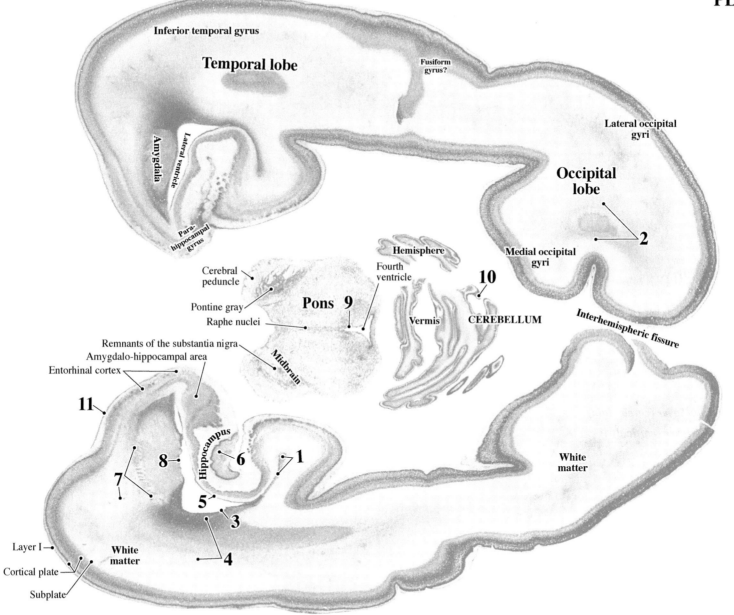

Inferior temporal gyrus

Temporal lobe

Fusiform gyrus?

Lateral occipital gyri

Occipital lobe

Amygdala

Lateral ventricle

Para-hippocampal gyrus

Medial occipital gyri

Hemisphere

Fourth ventricle

Cerebral peduncle

Pons **9**

10

Pontine gray

Raphe nuclei

Vermis

CEREBELLUM

Interhemispheric fissure

Remnants of the substantia nigra

Amygdalo-hippocampal area

Entorhinal cortex

Midbrain

11

Hippocampus

8

6

1

7

5

3

White matter

Layer I

White matter

4

Cortical plate

Subplate

PLATE 70A
CR 260 mm
GW 30, Y187-65
Horizontal
Section 941

See detail of the brain core
in Plates 92A and B.

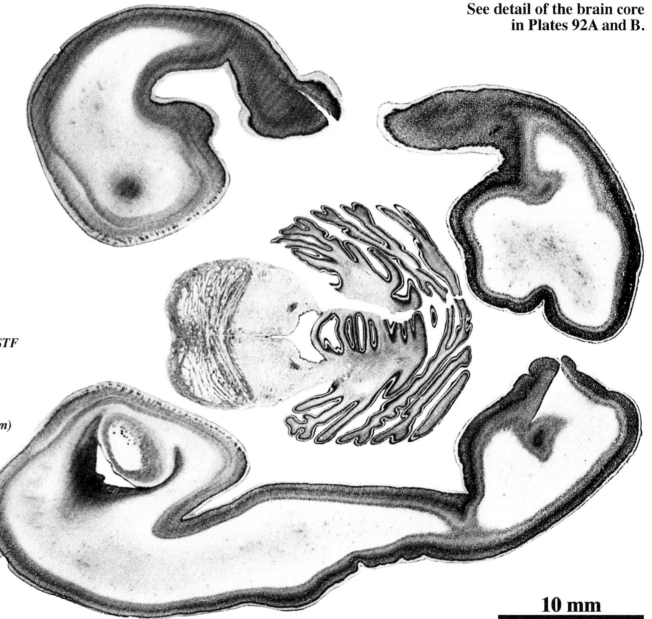

Remnants
of the germinal matrix,
migratory streams,
and transitional fields

1 *Parahippocampal NEP, SVZ, and STF*

2 *Temporal NEP, SVZ, and STF*

3 *Alvear GEP*

4 *Pontine G/EP*

5 *External germinal layer (cerebellum)*

6 *Subpial granular layer (cortical)*

GEP - Glioepithelium
G/EP - Glioepithelium/ependyma
NEP - Neuroepithelium
STF - Stratified transitional field
SVZ - Subventricular zone

10 mm

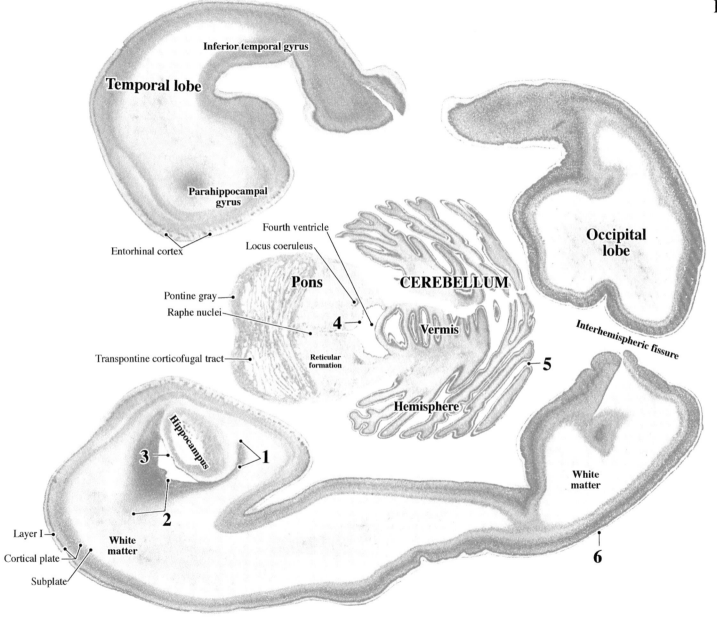

Inferior temporal gyrus

Temporal lobe

Parahippocampal gyrus

Fourth ventricle

Locus coeruleus

Entorhinal cortex

Pontine gray

Pons

CEREBELLUM

Occipital lobe

Raphe nuclei

4

Vermis

Interhemispheric fissure

Transpontine corticofugal tract

Reticular formation

5

Hemisphere

Hippocampus

3

1

2

White matter

Layer I

Cortical plate

White matter

Subplate

6

PLATE 71A
CR 260 mm
GW 30, Y187-65
Horizontal
Section 1041

See detail of the brain core
in Plates 93A and B.

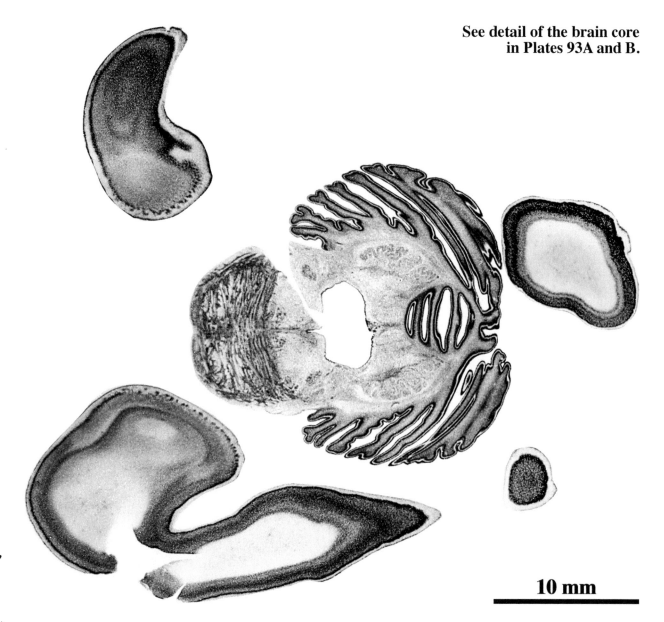

Remnants of the germinal matrix,
migratory streams,
and transitional fields

1 *Pontine G/EP*

2 *External germinal layer (cerebellum)*

3 *Subpial granular layer (cortical)*

G/EP - Glioepithelium/ependyma
STF - Stratified transitional field

10 mm

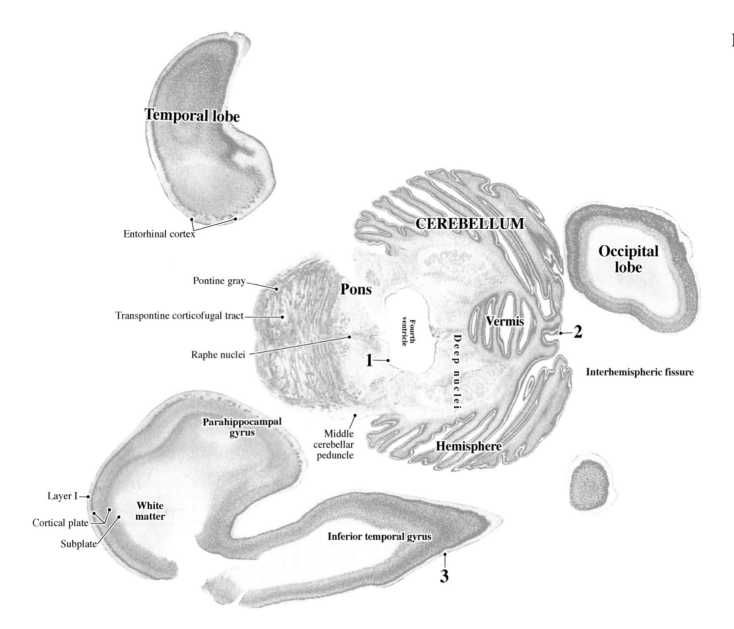

Temporal lobe

Entorhinal cortex

CEREBELLUM

Occipital lobe

Pontine gray

Pons

Transpontine corticofugal tract

Fourth ventricle

Vermis

Deep nuclei

2

Raphe nuclei

1

Interhemispheric fissure

Parahippocampal gyrus

Middle cerebellar peduncle

Hemisphere

Layer I

White matter

Cortical plate

Subplate

Inferior temporal gyrus

3

PLATE 72A
CR 260 mm
GW 30, Y187-65
Horizontal
Section 1081

See detail of the brain core
in Plates 94A and B.

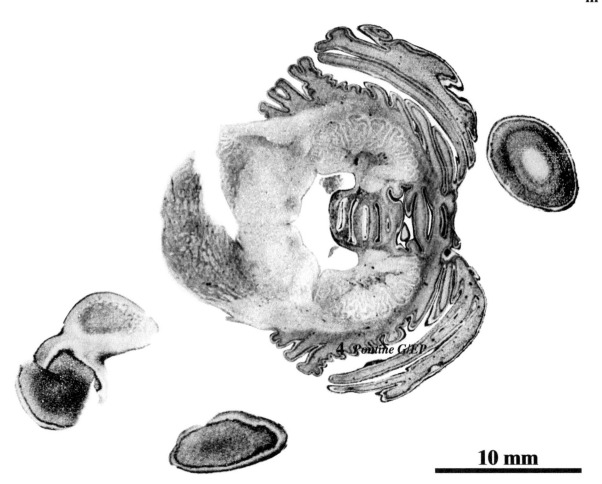

4 Pontine G/EP

10 mm

Remnants of the germinal matrix

1 *Pontine glioepithelium/ependyma*

2 *External germinal layer (cerebellum)*

3 *Subpial granular layer (cortical)*

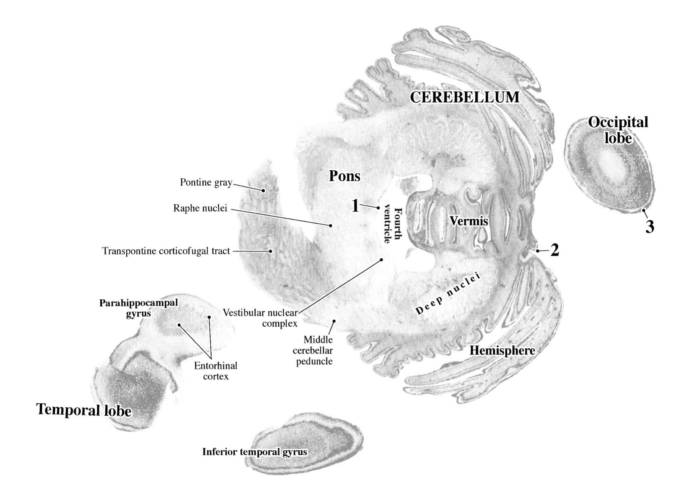

CEREBELLUM

Occipital
lobe

Pontine gray

Pons

Raphe nuclei

1

Fourth
ventricle

Vermis

3

Transpontine corticofugal tract

2

Deep nuclei

Parahippocampal
gyrus

Vestibular nuclear
complex

Middle
cerebellar
peduncle

Hemisphere

Entorhinal
cortex

Temporal lobe

Inferior temporal gyrus

PLATE 73A
CR 260 mm
GW 30, Y187-65
Horizontal
Section 1121

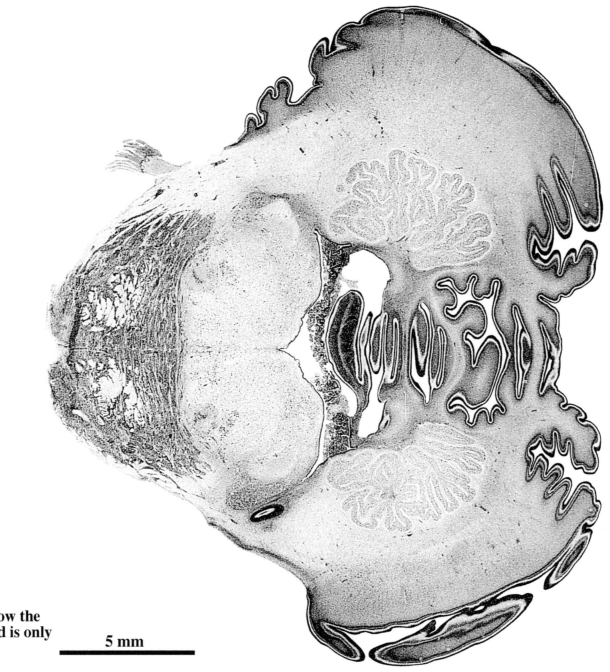

Section 1121 is below the
cerebral cortex and is only
shown at high
magnification.

5 mm

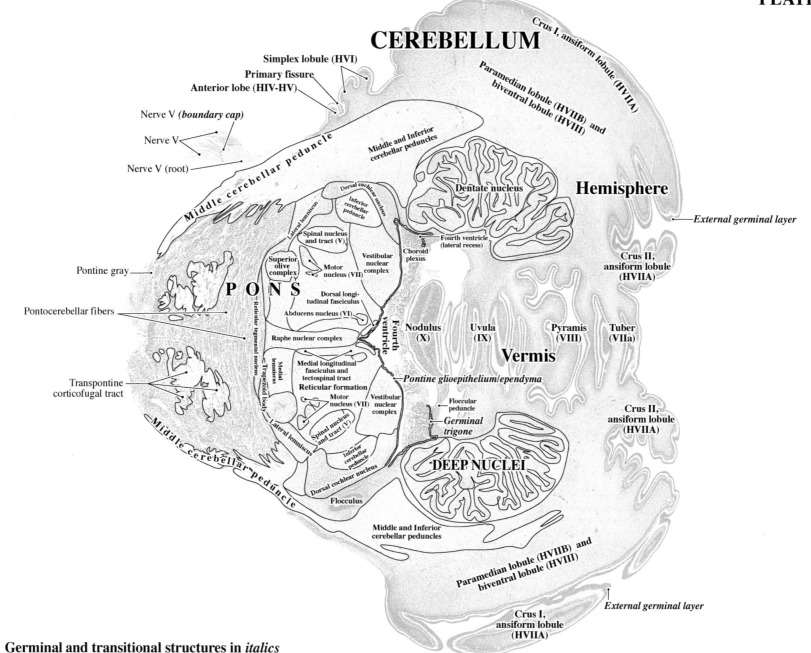

149

CEREBELLUM

Crus I, ansiform lobule (HVIIA)

Simplex lobule (HVI)
Primary fissure
Anterior lobe (HIV-HV)

Paramedian lobule (HVIIB) and
biventral lobule (HVIII)

Nerve V (*boundary cap*)

Nerve V

Nerve V (root)

Middle cerebellar peduncle

Middle and Inferior
cerebellar peduncles

Dorsal cochlear nucleus

Inferior
cerebellar
peduncle

Dentate nucleus

Hemisphere

Lateral lemniscus

Spinal nucleus
and tract (V)

Superior
olive
complex

Motor
nucleus (VII)

Vestibular
nuclear
complex

Fourth ventricle
(lateral recess)

Choroid
plexus

External germinal layer

Crus II,
ansiform lobule
(HVIIA)

Pontine gray

P O N S

Dorsal longi-
tudinal fasciculus

Abducens nucleus (VI)

Reticular tegmental nucleus

Raphe nuclear complex

Pontocerebellar fibers

Trapezoid body

**Fourth
ventricle**

**Nodulus
(X)**

**Uvula
(IX)**

**Pyramis
(VIII)**

**Tuber
(VIIa)**

Medial longitudinal
fasciculus and
tectospinal tract
Reticular formation

Medial
lemniscus

Motor
nucleus (VII)

Vestibular
nuclear
complex

Pontine glioepithelium/ependyma

Vermis

Floccular
peduncle

Transpontine
corticofugal tract

Lateral lemniscus

Spinal nucleus
and tract (V)

*Germinal
trigone*

Crus II,
ansiform lobule
(HVIIA)

Inferior
cerebellar
peduncle

DEEP NUCLEI

Middle cerebellar peduncle

Dorsal cochlear nucleus

Flocculus

Middle and Inferior
cerebellar peduncles

Paramedian lobule (HVIIB) and
biventral lobule (HVIII)

External germinal layer

Crus I,
ansiform lobule
(HVIIA)

Germinal and transitional structures in *italics*

150

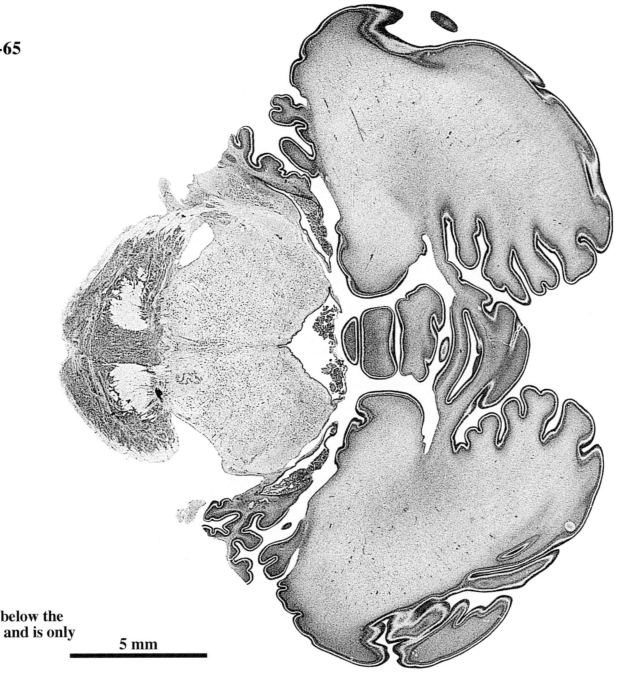

Section 1161 is below the
cerebral cortex and is only
shown at high
magnification.

5 mm

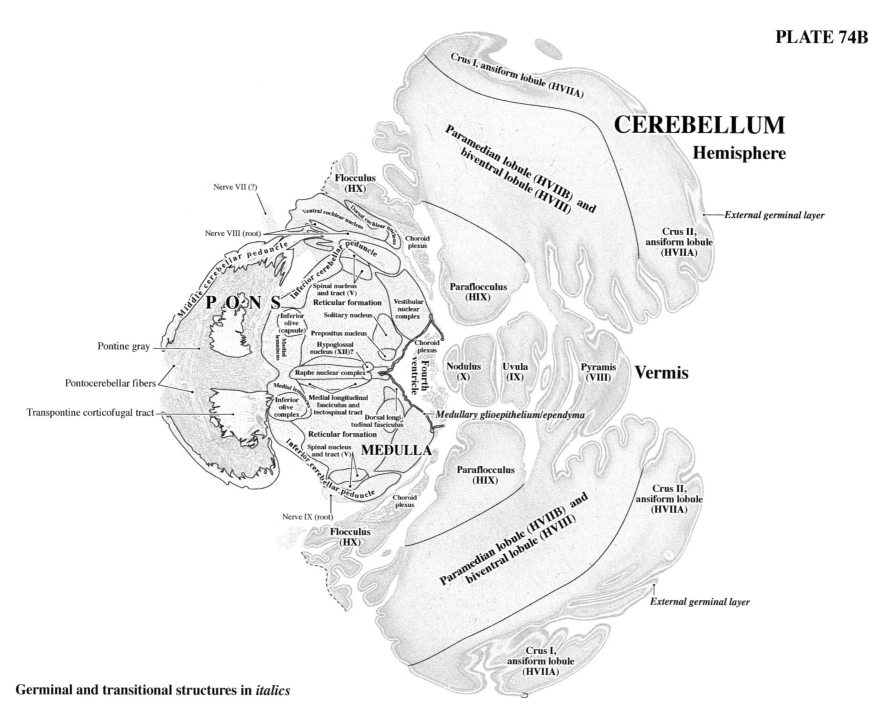

Crus I, ansiform lobule (HVIIA)

CEREBELLUM

Hemisphere

Paramedian lobule (HVIIB) and biventral lobule (HVIII)

Flocculus (HX)

Nerve VII (?)

Ventral cochlear nucleus

Dorsal cochlear nucleus

Nerve VIII (root)

Choroid plexus

External germinal layer

Crus II, ansiform lobule (HVIIA)

Middle cerebellar peduncle

Inferior cerebellar peduncle

Spinal nucleus and tract (V)

Reticular formation

Vestibular nuclear complex

Paraflocculus (HIX)

P O N S

Inferior olive (capsule)

Solitary nucleus

Medial lemniscus

Prepositus nucleus

Hypoglossal nucleus (XII)?

Choroid plexus

Pontine gray

Raphe nuclear complex

Fourth ventricle

Nodulus (X)

Uvula (IX)

Pyramis (VIII)

Vermis

Pontocerebellar fibers

Medial lemniscus

Inferior olive complex

Medial longitudinal fasciculus and tectospinal tract

Dorsal longitudinal fasciculus

Medullary glioepithelium/ependyma

Transpontine corticofugal tract

Reticular formation

MEDULLA

Spinal nucleus and tract (V)

Paraflocculus (HIX)

Inferior cerebellar peduncle

Crus II, ansiform lobule (HVIIA)

Choroid plexus

Nerve IX (root)

Flocculus (HX)

Paramedian lobule (HVIIB) and biventral lobule (HVIII)

External germinal layer

Crus I, ansiform lobule (HVIIA)

Germinal and transitional structures in *italics*

PLATE 75A
CR 260 mm
GW 30, Y187-65
Horizontal
Section 1161

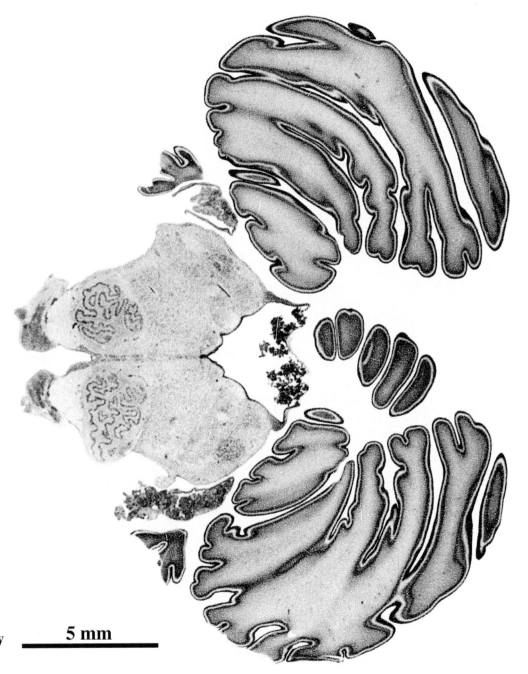

Section 1201 is below the
cerebral cortex and is only
shown at high
magnification.

5 mm

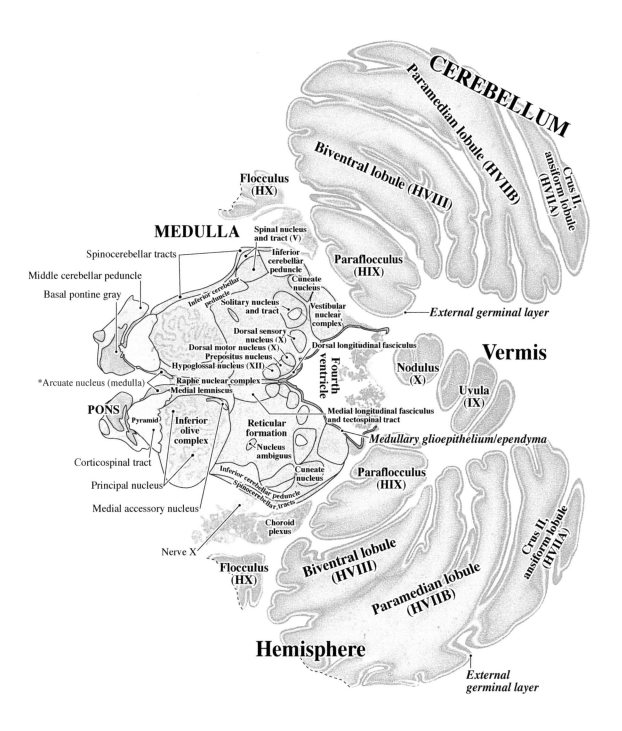

CEREBELLUM

Paramedian lobule (HVIIB)

Biventral lobule (HVIII)

Crus II, ansiform lobule (HVIIA)

Flocculus (HX)

MEDULLA

Spinal nucleus and tract (V)

Spinocerebellar tracts

Inferior cerebellar peduncle

Middle cerebellar peduncle

Inferior cerebellar peduncle

Basal pontine gray

Cuneate nucleus

Paraflocculus (HIX)

Solitary nucleus and tract

Vestibular nuclear complex

External germinal layer

Dorsal sensory nucleus (X)

Dorsal motor nucleus (X)

Dorsal longitudinal fasciculus

Vermis

Prepositus nucleus

Hypoglossal nucleus (XII)

Fourth ventricle

Nodulus (X)

*Arcuate nucleus (medulla)

Raphe nuclear complex

Medial lemniscus

Uvula (IX)

PONS

Pyramid

Inferior olive complex

Reticular formation

Medial longitudinal fasciculus and tectospinal tract

Medullary glioepithelium/ependyma

Nucleus ambiguus

Corticospinal tract

Cuneate nucleus

Paraflocculus (HIX)

Principal nucleus

Inferior cerebellar peduncle

Medial accessory nucleus

Spinocerebellar tracts

Crus II, ansiform lobule (HVIIA)

Choroid plexus

Nerve X

Flocculus (HX)

Biventral lobule (HVIII)

Paramedian lobule (HVIIB)

Hemisphere

External germinal layer

154

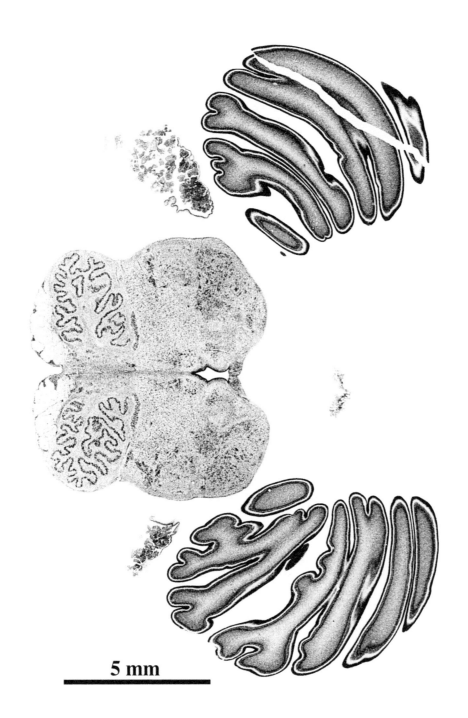

Section 1241 is below the
cerebral cortex and is only
shown at high
magnification.

5 mm

CEREBELLUM

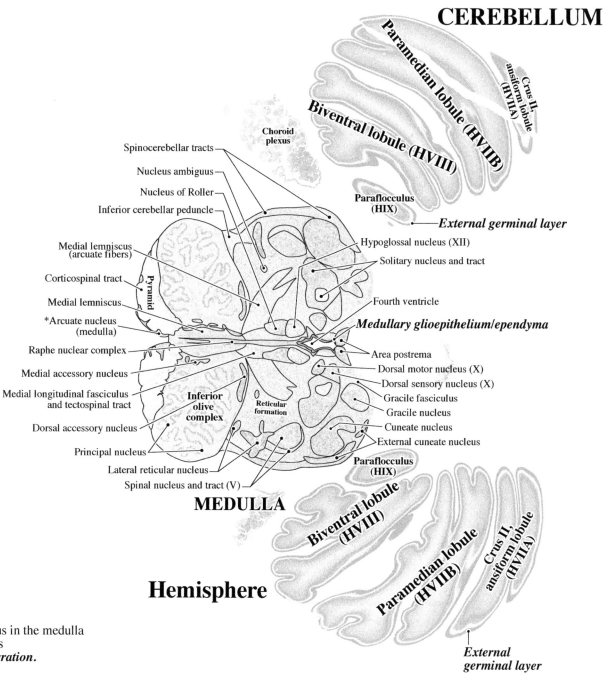

Paramedian lobule (HVIIB)

Crus II, ansiform lobule (HVIIA)

Biventral lobule (HVIII)

Paraflocculus (HIX)

Choroid plexus

Spinocerebellar tracts

Nucleus ambiguus

Nucleus of Roller

Inferior cerebellar peduncle

Medial lemniscus (arcuate fibers)

Corticospinal tract

Medial lemniscus

*Arcuate nucleus (medulla)

Raphe nuclear complex

Medial accessory nucleus

Medial longitudinal fasciculus and tectospinal tract

Dorsal accessory nucleus

Principal nucleus

Lateral reticular nucleus

Spinal nucleus and tract (V)

Pyramid

Inferior olive complex

Reticular formation

External germinal layer

Hypoglossal nucleus (XII)

Solitary nucleus and tract

Fourth ventricle

Medullary glioepithelium/ependyma

Area postrema

Dorsal motor nucleus (X)

Dorsal sensory nucleus (X)

Gracile fasciculus

Gracile nucleus

Cuneate nucleus

External cuneate nucleus

Paraflocculus (HIX)

MEDULLA

Biventral lobule (HVIII)

Paramedian lobule (HVIIB)

Crus II, ansiform lobule (HVIIA)

Hemisphere

External germinal layer

*The arcuate nucleus in the medulla
may contain neurons
from the *Raphe migration.*

PLATE 77A
CR 260 mm
GW 30, Y187-65
Horizontal
Section 1281

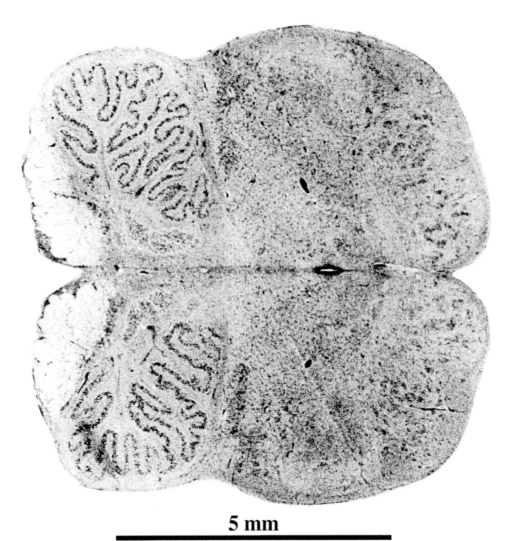

5 mm

Section 1281 is below the
cerebral cortex and is only
shown at high
magnification.

LOWER MEDULLA

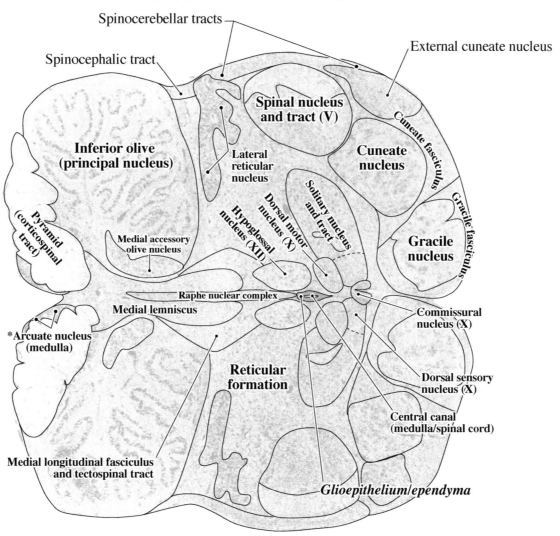

Spinocerebellar tracts

External cuneate nucleus

Spinocephalic tract

Spinal nucleus and tract (V)

Cuneate fasciculus

Inferior olive (principal nucleus)

Lateral reticular nucleus

Cuneate nucleus

Gracile fasciculus

Solitary nucleus and tract

Dorsal motor nucleus (X)

Pyramid (corticospinal tract)

Hypoglossal nucleus (XII)

Gracile nucleus

Medial accessory olive nucleus

Raphe nuclear complex

Commissural nucleus (X)

Medial lemniscus

*Arcuate nucleus (medulla)

Dorsal sensory nucleus (X)

Reticular formation

Central canal (medulla/spinal cord)

Medial longitudinal fasciculus and tectospinal tract

Glioepithelium/ependyma

*The arcuate nucleus in the medulla
may contain neurons
from the ***Raphe migration.***

Germinal and transitional structures in *italics*

PLATE 78A
CR 260 mm
GW 30, Y187-65
Horizontal
Section 1351

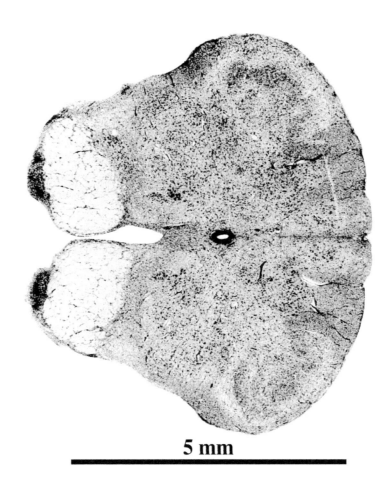

5 mm

Section 1351 is below the
cerebral cortex and is only
shown at high
magnification.

MEDULLA/
SPINAL CORD

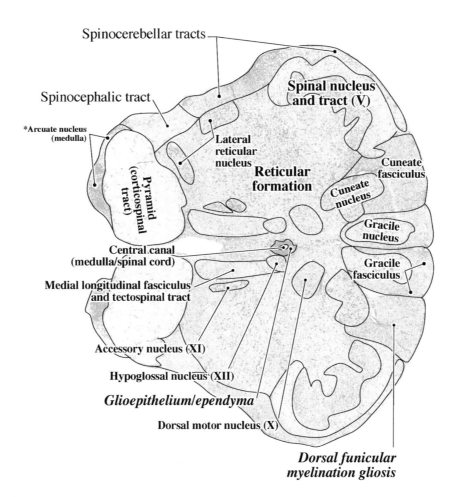

Spinocerebellar tracts

Spinocephalic tract

Spinal nucleus
and tract (V)

*Arcuate nucleus
(medulla)

Lateral
reticular
nucleus

Cuneate
fasciculus

Reticular
formation

Cuneate
nucleus

Pyramid
(corticospinal
tract)

Gracile
nucleus

Central canal
(medulla/spinal cord)

Gracile
fasciculus

Medial longitudinal fasciculus
and tectospinal tract

Accessory nucleus (XI)

Hypoglossal nucleus (XII)

Glioepithelium/ependyma

Dorsal motor nucleus (X)

*Dorsal funicular
myelination gliosis*

*The arcuate nucleus in the medulla
may contain neurons
from the ***Raphe migration.***

Germinal and transitional structures in *italics*

PLATE 79A
CR 260 mm
GW 30, Y187-65
Horizontal
Section 1391

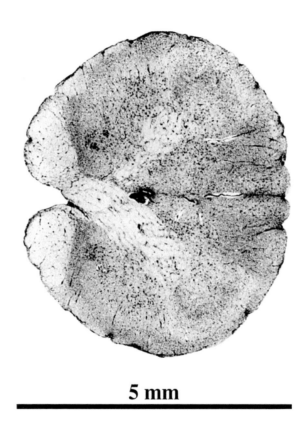

5 mm

Section 1391 is below the cerebral cortex and is only shown at high magnification.

MEDULLA/
SPINAL CORD

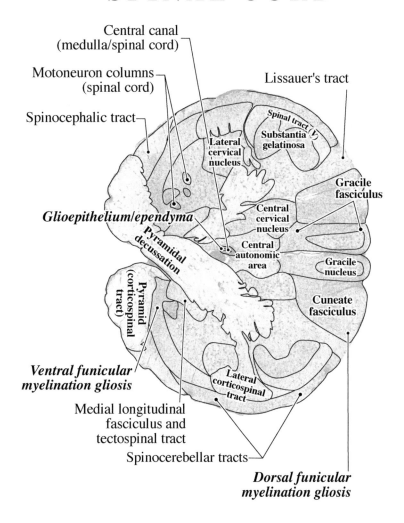

Central canal
(medulla/spinal cord)

Motoneuron columns
(spinal cord)

Spinocephalic tract

Lissauer's tract

Spinal tract

Substantia
gelatinosa

Lateral
cervical
nucleus

Glioepithelium/ependyma

Gracile
fasciculus

Central
cervical
nucleus

*Pyramidal
decussation*

Central
autonomic
area

Gracile
nucleus

*Pyramid
(corticospinal
tract)*

Cuneate
fasciculus

*Ventral funicular
myelination gliosis*

Lateral
corticospinal
tract

Medial longitudinal
fasciculus and
tectospinal tract

Spinocerebellar tracts

*Dorsal funicular
myelination gliosis*

Germinal and transitional structures in *italics*

PLATE 80A
CR 260 mm, GW 30, Y187-65, Horizontal, Section 441

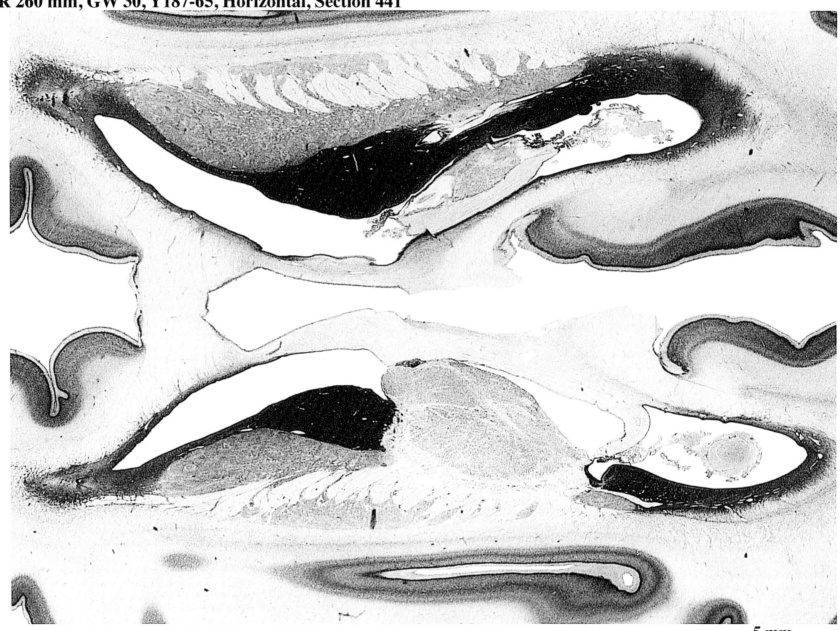

See the entire Section 441 in Plates 58A and B.

5 mm

Germinal and transitional structures in *italics*

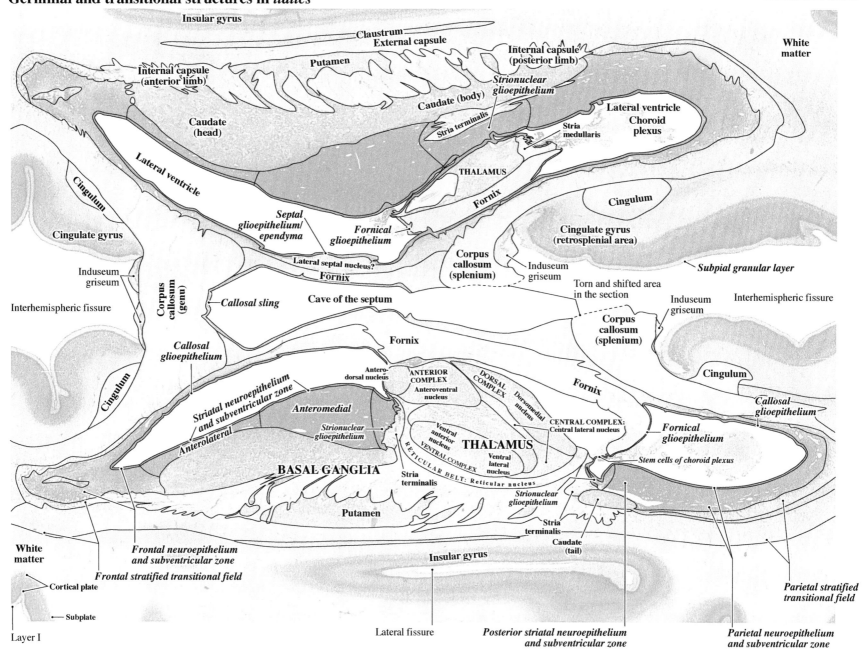

Insular gyrus

Claustrum

External capsule

Internal capsule
(posterior limb)

White
matter

Putamen

Internal capsule
(anterior limb)

*Strionuclear
glioepithelium*

Caudate (body)

Lateral ventricle
Choroid
plexus

Caudate
(head)

Stria terminalis

Stria
medullaris

Lateral ventricle

THALAMUS

Cingulum

Fornix

Cingulum

Cingulate gyrus
(retrosplenial area)

Cingulate gyrus

*Septal
glioepithelium/
ependyma*

*Fornical
glioepithelium*

Induseum
griseum

Subpial granular layer

Induseum
griseum

Lateral septal nucleus?

Corpus
callosum
(splenium)

Interhemispheric fissure

Fornix

Torn and shifted area
in the section

Induseum
griseum

Interhemispheric fissure

Corpus
callosum
(genu)

Cave of the septum

Corpus
callosum
(splenium)

Callosal sling

Fornix

Cingulum

*Callosal
glioepithelium*

Fornix

Antero-
dorsal nucleus

ANTERIOR
COMPLEX

DORSAL
COMPLEX

*Callosal-
glioepithelium*

*Striatal neuroepithelium
and subventricular zone*

Anteroventral
nucleus

Anteromedial

Dorsomedial
nucleus

Cingulum

*Strionuclear
glioepithelium*

Ventral
anterior
nucleus

CENTRAL COMPLEX:
Central lateral nucleus

*Fornical
glioepithelium*

Anterolateral

VENTRAL COMPLEX

THALAMUS

BASAL GANGLIA

RETICULAR BELT: Reticular nucleus

Ventral
lateral
nucleus

Stem cells of choroid plexus

Stria
terminalis

*Strionuclear
glioepithelium*

Putamen

Stria
terminalis

Caudate
(tail)

White
matter

Insular gyrus

*Parietal stratified
transitional field*

Cortical plate

*Frontal neuroepithelium
and subventricular zone*

Frontal stratified transitional field

Subplate

Layer I

Lateral fissure

*Posterior striatal neuroepithelium
and subventricular zone*

*Parietal neuroepithelium
and subventricular zone*

PLATE 81A
CR 260 mm, GW 30, Y187-65, Horizontal, Section 521

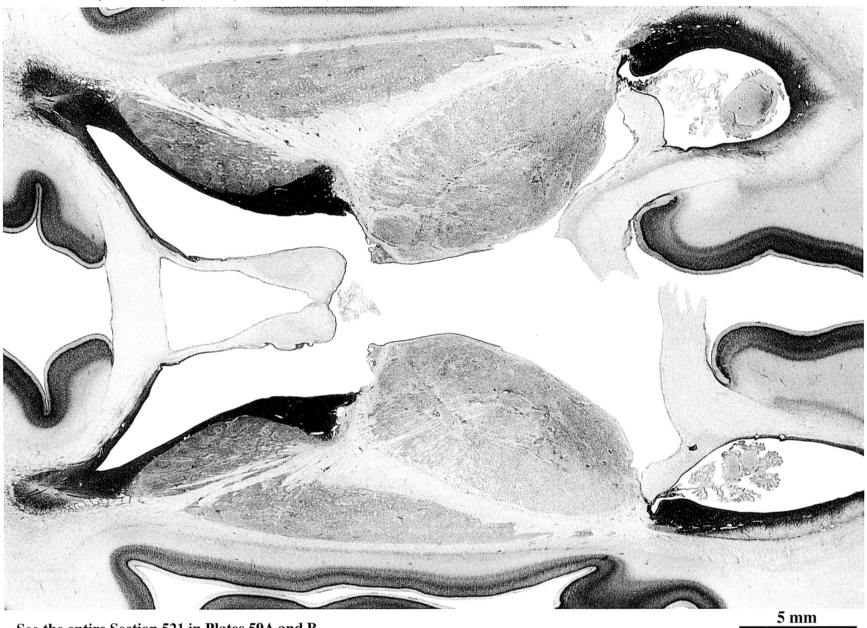

5 mm

See the entire Section 521 in Plates 59A and B.

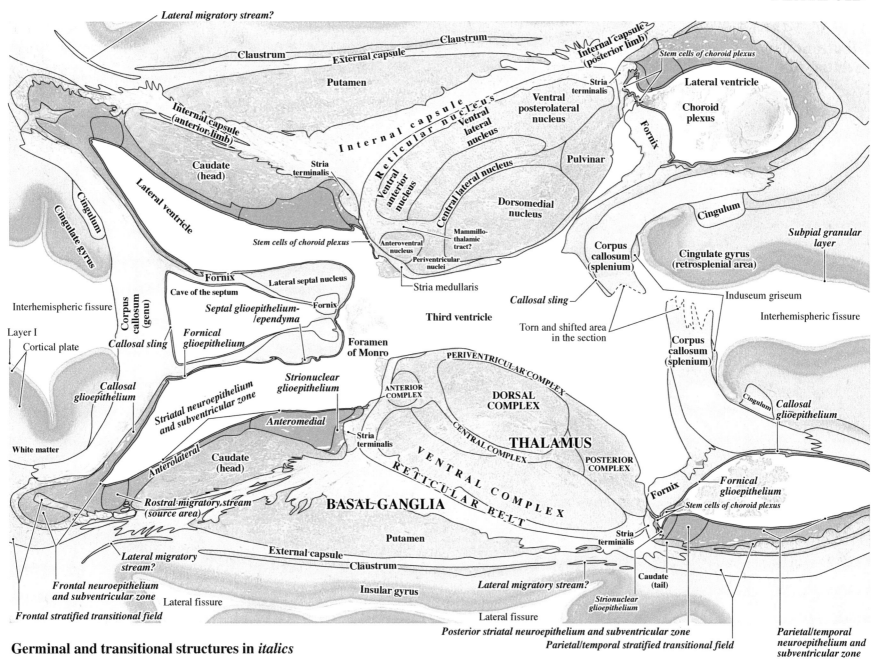

Germinal and transitional structures in *italics*

PLATE 82A
CR 260 mm
GW 30, Y187-65
Horizontal
Section 621

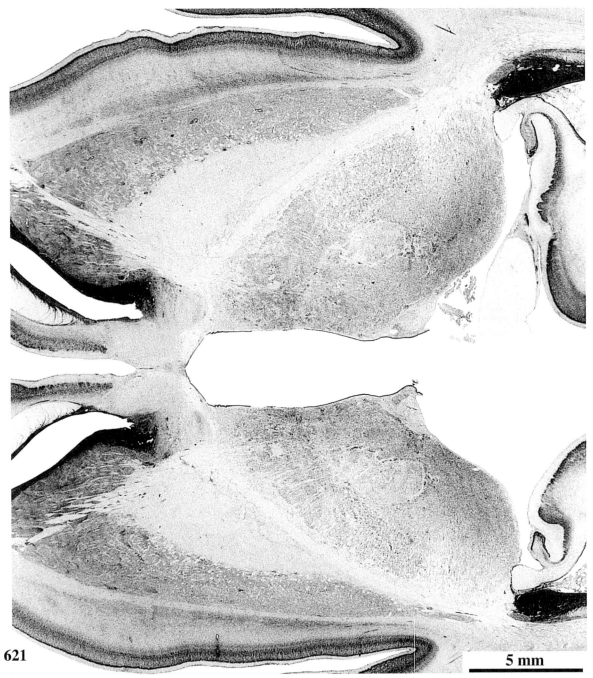

See the entire Section 621
in Plates 60A and B.

5 mm

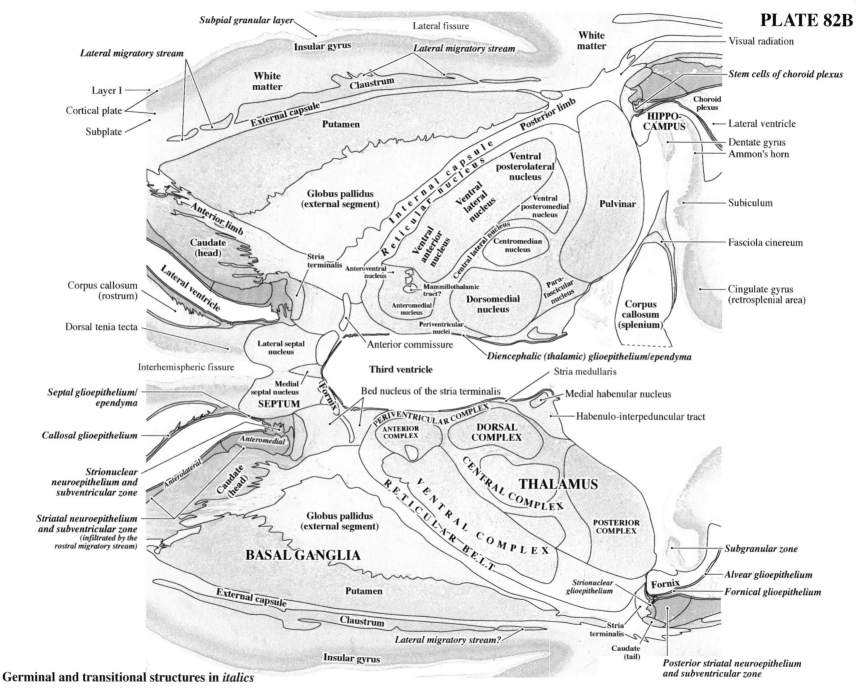

Germinal and transitional structures in *italics*

PLATE 83A
CR 260 mm
GW 30, Y187-65
Horizontal
Section 671

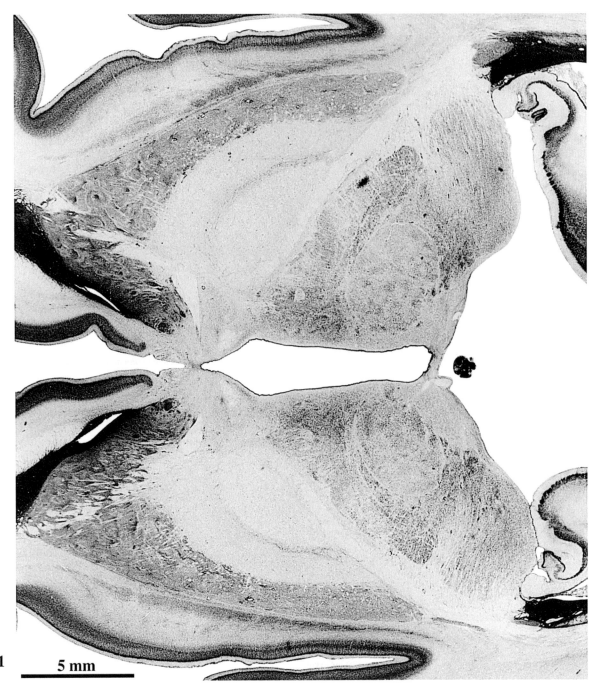

See the entire Section 671
in Plates 61A and B.

5 mm

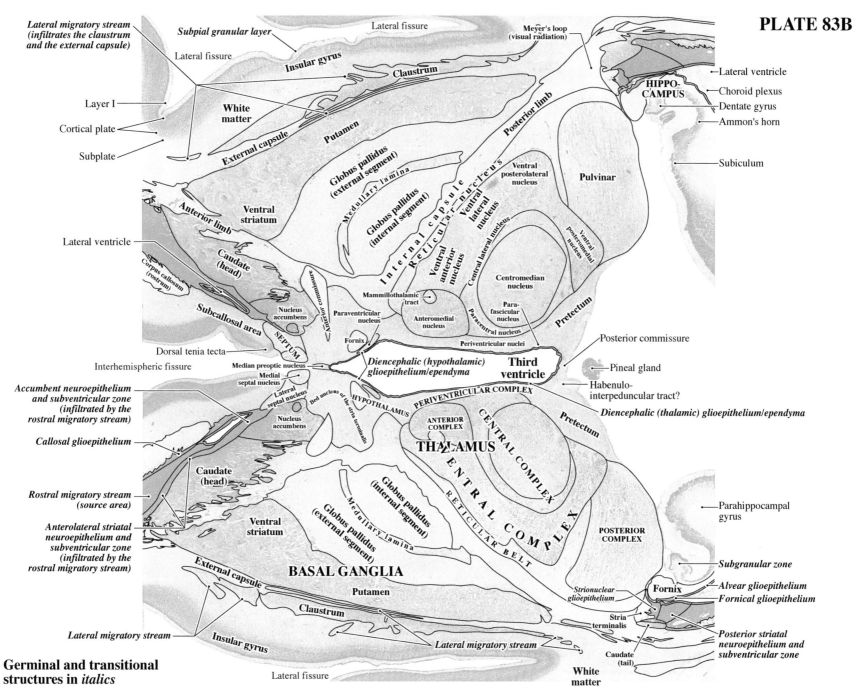

Lateral migratory stream (infiltrates the claustrum and the external capsule)

Lateral fissure

Subpial granular layer

Lateral fissure

Insular gyrus

Claustrum

Meyer's loop (visual radiation)

Lateral ventricle

HIPPO-CAMPUS

Choroid plexus

Dentate gyrus

Ammon's horn

Layer I

Cortical plate

Subplate

White matter

Putamen

Posterior limb

External capsule

Subiculum

Globus pallidus (external segment)

Medullary lamina

Globus pallidus (internal segment)

Internal capsule

Reticular nucleus

Ventral posterolateral nucleus

Pulvinar

Ventral lateral nucleus

Ventral posteromedial nucleus

Anterior limb

Ventral striatum

Lateral ventricle

Corpus callosum (rostrum)

Caudate (head)

Anterior commissure

Ventral anterior nucleus

Central lateral nucleus

Centromedian nucleus

Mammillothalamic tract

Para-fascicular nucleus

Pretectum

Nucleus accumbens

Paraventricular nucleus

Anteromedial nucleus

Paracentral nucleus

Subcallosal area

Fornix

Periventricular nuclei

Posterior commissure

SEPTUM

Dorsal tenia tecta

Median preoptic nucleus

Diencephalic (hypothalamic) glioepithelium/ependyma

Third ventricle

Pineal gland

Interhemispheric fissure

Medial septal nucleus

Habenulo-interpeduncular tract?

Accumbent neuroepithelium and subventricular zone (infiltrated by the rostral migratory stream)

Lateral septal nucleus

Bed nucleus of the stria terminalis

HYPOTHALAMUS

PERIVENTRICULAR COMPLEX

Diencephalic (thalamic) glioepithelium/ependyma

Nucleus accumbens

ANTERIOR COMPLEX

CENTRAL COMPLEX

Pretectum

Callosal glioepithelium

THALAMUS

CENTRAL COMPLEX

Caudate (head)

Rostral migratory stream (source area)

Globus pallidus (internal segment)

RETICULAR BELT

Parahippocampal gyrus

Anterolateral striatal neuroepithelium and subventricular zone (infiltrated by the rostral migratory stream)

Ventral striatum

Medullary lamina

Globus pallidus (external segment)

POSTERIOR COMPLEX

Subgranular zone

External capsule

BASAL GANGLIA

Putamen

Fornix

Alvear glioepithelium

Fornical glioepithelium

Claustrum

Strionuclear glioepithelium

Stria terminalis

Lateral migratory stream

Insular gyrus

Lateral migratory stream

Posterior striatal neuroepithelium and subventricular zone

Caudate (tail)

Germinal and transitional structures in *italics*

Lateral fissure

White matter

PLATE 84A
CR 260 mm
GW 30, Y187-65
Horizontal
Section 701

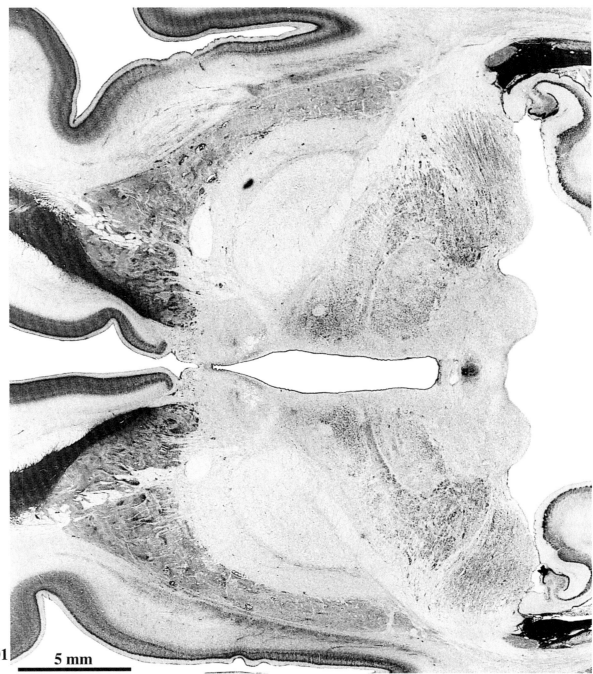

See the entire Section 701
in Plates 62A and B.

5 mm

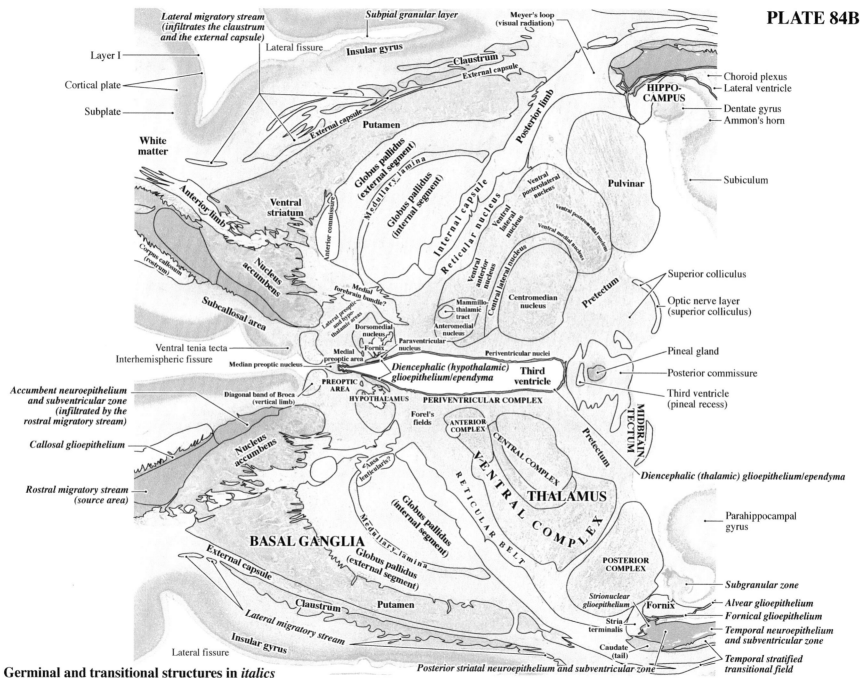

Lateral migratory stream
(infiltrates the claustrum
and the external capsule)

Subpial granular layer

Meyer's loop
(visual radiation)

Layer I

Lateral fissure

Insular gyrus

Claustrum

External capsule

Choroid plexus

Lateral ventricle

HIPPO-
CAMPUS

Cortical plate

Dentate gyrus

Ammon's horn

Subplate

Putamen

External capsule

White
matter

Globus pallidus
(external segment)

Medullary lamina

Globus pallidus
(internal segment)

Posterior limb

Internal capsule

Reticular nucleus

Ventral
posterolateral
nucleus

Pulvinar

Subiculum

Anterior limb

Ventral
striatum

Anterior commissure

Ventral posteromedial nucleus

Ventral
lateral
nucleus

Corpus callosum
(rostrum)

Nucleus
accumbens

Ventral
anterior
nucleus

Central lateral nucleus

Ventral medial nucleus

Medial
forebrain
bundle?

Mammillo-
thalamic
tract

Centromedian
nucleus

Superior colliculus

Subcallosal area

Lateral preoptic
and hypo-
thalamic areas

Anteromedial
nucleus

Pretectum

Optic nerve layer
(superior colliculus)

Dorsomedial
nucleus

Ventral tenia tecta

Fornix

Paraventricular
nucleus

Interhemispheric fissure

Median preoptic nucleus

Medial
preoptic area

Periventricular nuclei

Pineal gland

Posterior commissure

Diencephalic (hypothalamic)
glioepithelium/ependyma

Third
ventricle

Third ventricle
(pineal recess)

Accumbent neuroepithelium
and subventricular zone
(infiltrated by the
rostral migratory stream)

PREOPTIC
AREA

Diagonal band of Broca
(vertical limb)

HYPOTHALAMUS

PERIVENTRICULAR COMPLEX

Forel's
fields

ANTERIOR
COMPLEX

CENTRAL COMPLEX

MIDBRAIN TECTUM

Pretectum

Diencephalic (thalamic) glioepithelium/ependyma

Callosal glioepithelium

Nucleus
accumbens

Ansa
lenticularis?

VENTRAL COMPLEX

RETICULAR BELT

THALAMUS

Rostral migratory stream
(source area)

Globus pallidus
(internal segment)

Parahippocampal
gyrus

BASAL GANGLIA

Medullary lamina

Globus pallidus
(external segment)

POSTERIOR
COMPLEX

External capsule

Subgranular zone

Claustrum

Putamen

Strionuclear
glioepithelium

Fornix

Alvear glioepithelium

Fornical glioepithelium

Lateral migratory stream

Stria
terminalis

Temporal neuroepithelium
and subventricular zone

Insular gyrus

Caudate
(tail)

Lateral fissure

Posterior striatal neuroepithelium and subventricular zone

Temporal stratified
transitional field

Germinal and transitional structures in *italics*

PLATE 85A
CR 260 mm
GW 30, Y187-65
Horizontal
Section 721

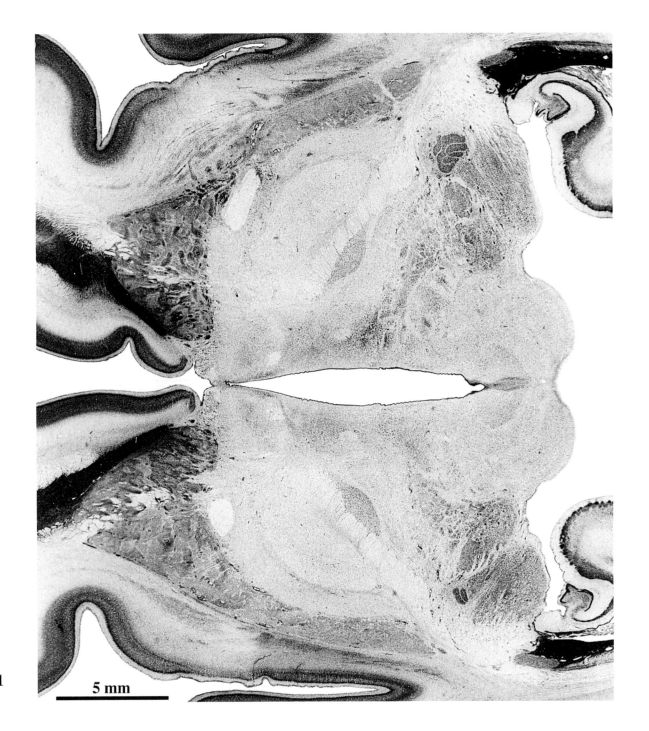

5 mm

See the entire Section 721
in Plates 63A and B.

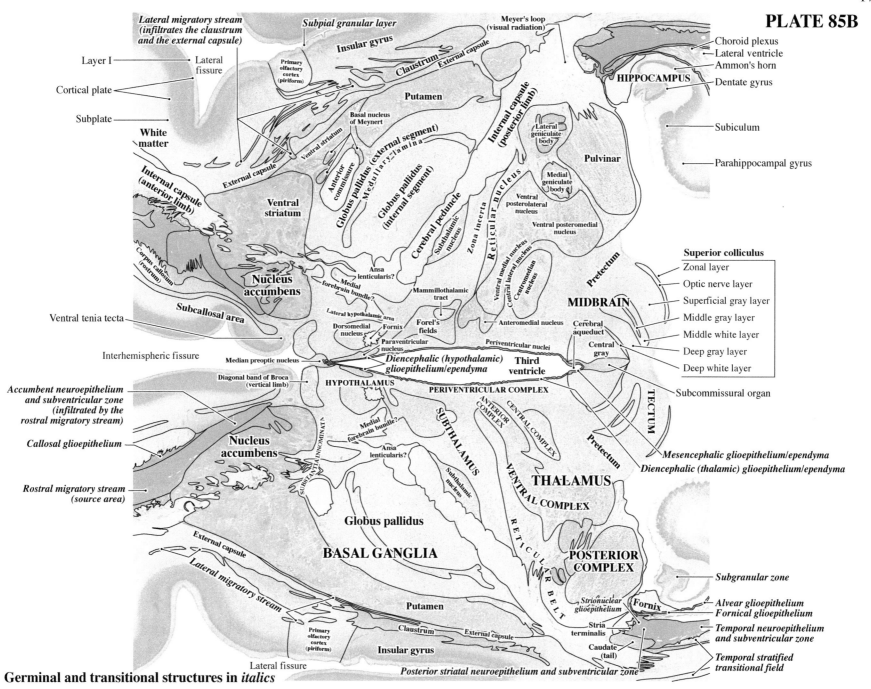

Lateral migratory stream (infiltrates the claustrum and the external capsule)

Subpial granular layer

Meyer's loop (visual radiation)

Choroid plexus

Lateral ventricle

Ammon's horn

Dentate gyrus

Layer I

Lateral fissure

Insular gyrus

Claustrum

External capsule

HIPPOCAMPUS

Cortical plate

Primary olfactory cortex (piriform)

Putamen

Internal capsule (posterior limb)

Subplate

Basal nucleus of Meynert

Lateral geniculate body

Subiculum

White matter

Ventral striatum

Pulvinar

Parahippocampal gyrus

Globus pallidus (external segment)

Medullary lamina

Anterior commissure

Medial geniculate body

Internal capsule (anterior limb)

External capsule

Globus pallidus (internal segment)

Cerebral peduncle

Zona incerta

Ventral posterolateral nucleus

Ventral posteromedial nucleus

Superior colliculus

Zonal layer

Ventral striatum

Subthalamic nucleus

Reticular nucleus

Optic nerve layer

Superficial gray layer

Corpus callosum (rostrum)

Nucleus accumbens

Medial forebrain bundle?

Ansa lenticularis?

Ventral medial nucleus

Central lateral nucleus

Centromedian nucleus

MIDBRAIN

Pretectum

Middle gray layer

Middle white layer

Ventral tenia tecta

Subcallosal area

Lateral hypothalamic area

Mammillothalamic tract

Forel's fields

Anteromedial nucleus

Cerebral aqueduct

Central gray

Deep gray layer

Deep white layer

Dorsomedial nucleus

Fornix

Paraventricular nucleus

Periventricular nuclei

Interhemispheric fissure

Median preoptic nucleus

Diencephalic (hypothalamic) glioepithelium/ependyma

Third ventricle

Subcommissural organ

Accumbent neuroepithelium and subventricular zone (infiltrated by the rostral migratory stream)

Diagonal band of Broca (vertical limb)

HYPOTHALAMUS

PERIVENTRICULAR COMPLEX

ANTERIOR COMPLEX

CENTRAL COMPLEX

TECTUM

Callosal glioepithelium

Nucleus accumbens

Medial forebrain bundle?

SUBTHALAMUS

VENTRAL COMPLEX

Pretectum

Mesencephalic glioepithelium/ependyma

Diencephalic (thalamic) glioepithelium/ependyma

Rostral migratory stream (source area)

Ansa lenticularis?

Subthalamic nucleus

THALAMUS

SUBSTANTIA INNOMINATA

Globus pallidus

BASAL GANGLIA

VENTRAL COMPLEX

RETICULAR BELT

POSTERIOR COMPLEX

Subgranular zone

Strionuclear glioepithelium

Fornix

Alvear glioepithelium

Fornical glioepithelium

Lateral migratory stream

External capsule

Putamen

Stria terminalis

Temporal neuroepithelium and subventricular zone

Primary olfactory cortex (piriform)

Claustrum

External capsule

Caudate (tail)

Temporal stratified transitional field

Insular gyrus

Lateral fissure

Posterior striatal neuroepithelium and subventricular zone

Germinal and transitional structures in *italics*

PLATE 86A
CR 260 mm
GW 30, Y187-65
Horizontal
Section 761

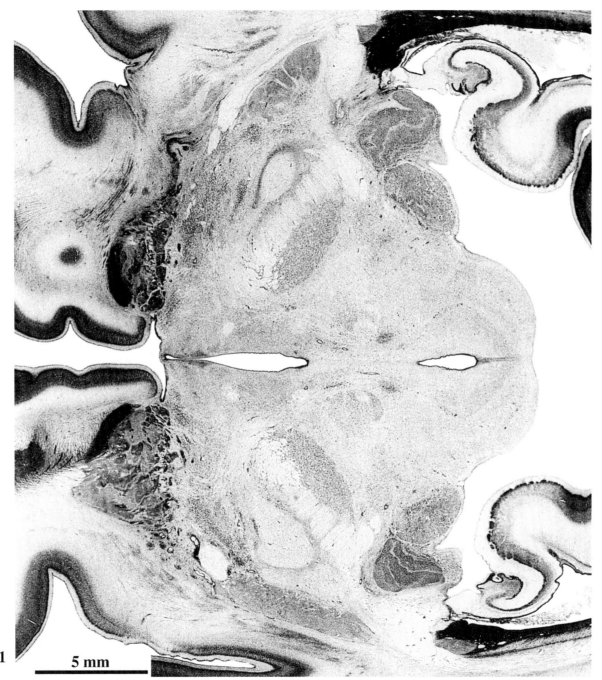

See the entire Section 761
in Plates 64A and B.

5 mm

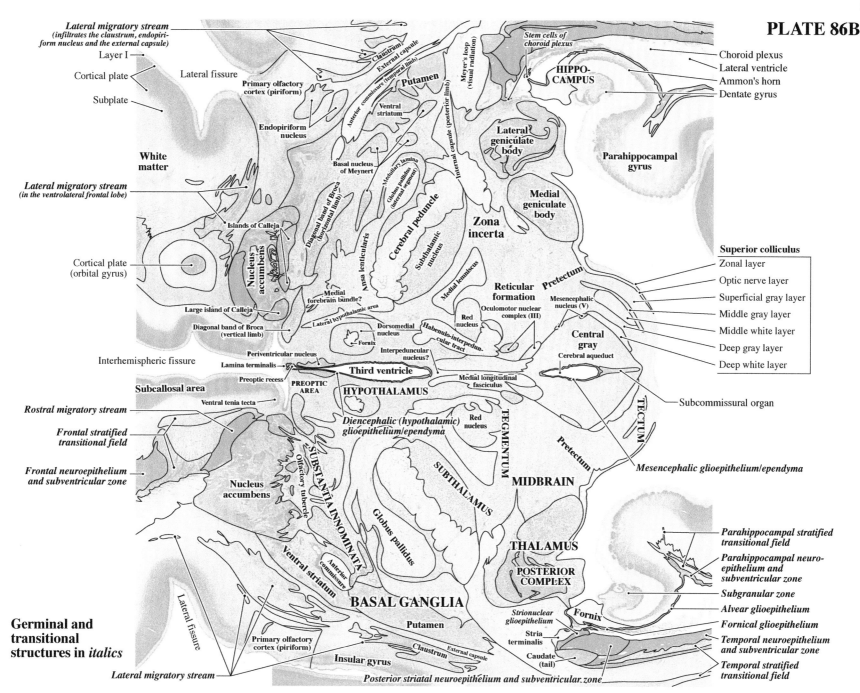

Lateral migratory stream
(infiltrates the claustrum, endopiri-
form nucleus and the external capsule)

Layer I

Cortical plate

Subplate

Lateral fissure

**White
matter**

Lateral migratory stream
(in the ventrolateral frontal lobe)

Cortical plate
(orbital gyrus)

Islands of Calleja

Large island of Calleja

Diagonal band of Broca
(vertical limb)

Interhemispheric fissure

Subcallosal area

Rostral migratory stream

*Frontal stratified
transitional field*

*Frontal neuroepithelium
and subventricular zone*

**Nucleus
accumbens**

**Germinal and
transitional
structures in** *italics*

Lateral migratory stream

Claustrum
External capsule

Primary olfactory
cortex (piriform)

Endopiriform
nucleus

Basal nucleus
of Meynert

Nucleus
accumbens

Medial
forebrain bundle?

Diagonal band of Broca
(horizontal limb)

Lateral hypothalamic area

Periventricular nucleus

Lamina terminalis

Preoptic recess

Ventral tenia tecta

Anterior commissure (temporal limb)

Putamen

Ventral
striatum

Medullary lamina

Globus pallidus
(lateral segment)

Ansa lenticularis

Internal capsule (posterior limb)

Meyer's loop
(visual radiation)

Stem cells of
choroid plexus

**HIPPO-
CAMPUS**

Choroid plexus

Lateral ventricle

Ammon's horn

Dentate gyrus

Lateral
geniculate
body

**Parahippocampal
gyrus**

Cerebral peduncle

Subthalamic
nucleus

**Zona
incerta**

**Medial
geniculate
body**

Medial lemniscus

**Reticular
formation**

Pretectum

Mesencephalic
nucleus (V)

Superior colliculus

Zonal layer

Optic nerve layer

Superficial gray layer

Middle gray layer

Middle white layer

Deep gray layer

Deep white layer

Oculomotor nuclear
complex (III)

Red
nucleus

Dorsomedial
nucleus

Habenulo-interpedun-
cular tract

Fornix

Interpeduncular
nucleus?

Third ventricle

HYPOTHALAMUS

**PREOPTIC
AREA**

*Diencephalic (hypothalamic)
glioepithelium/ependyma*

Medial longitudinal
fasciculus

**Central
gray**

Cerebral aqueduct

TECTUM

Pretectum

Subcommissural organ

Red
nucleus

TEGMENTUM

Mesencephalic glioepithelium/ependyma

SUBTHALAMUS

MIDBRAIN

SUBSTANTIA INNOMINATA

Olfactory tubercle

**Globus
pallidus**

Anterior
commissure

THALAMUS

**POSTERIOR
COMPLEX**

Ventral striatum

Lateral fissure

Primary olfactory
cortex (piriform)

Insular gyrus

Claustrum
External capsule

BASAL GANGLIA

Putamen

*Strionuclear
glioepithelium*

Fornix

Stria
terminalis

Caudate
(tail)

Posterior striatal neuroepithelium and subventricular zone

*Parahippocampal stratified
transitional field*

*Parahippocampal neuro-
epithelium and
subventricular zone*

Subgranular zone

Alvear glioepithelium

Fornical glioepithelium

*Temporal neuroepithelium
and subventricular zone*

*Temporal stratified
transitional field*

PLATE 87A
CR 260 mm
GW 30, Y187-65
Horizontal
Section 801

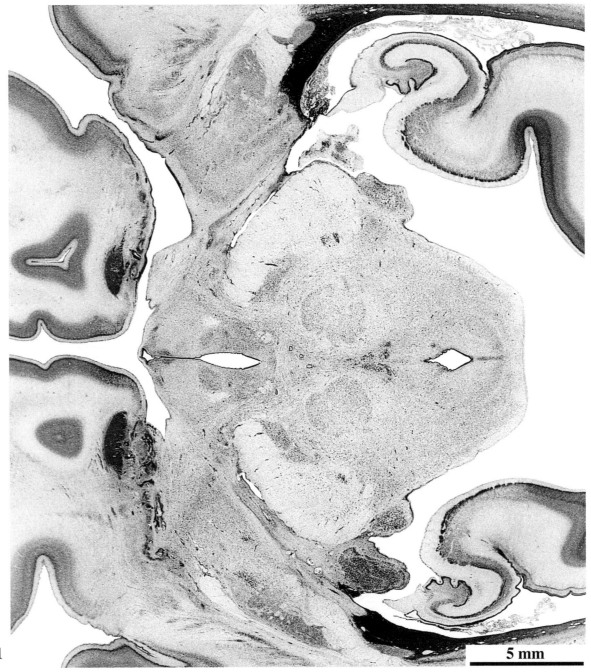

5 mm

**See the entire Section 801
in Plates 65A and B.**

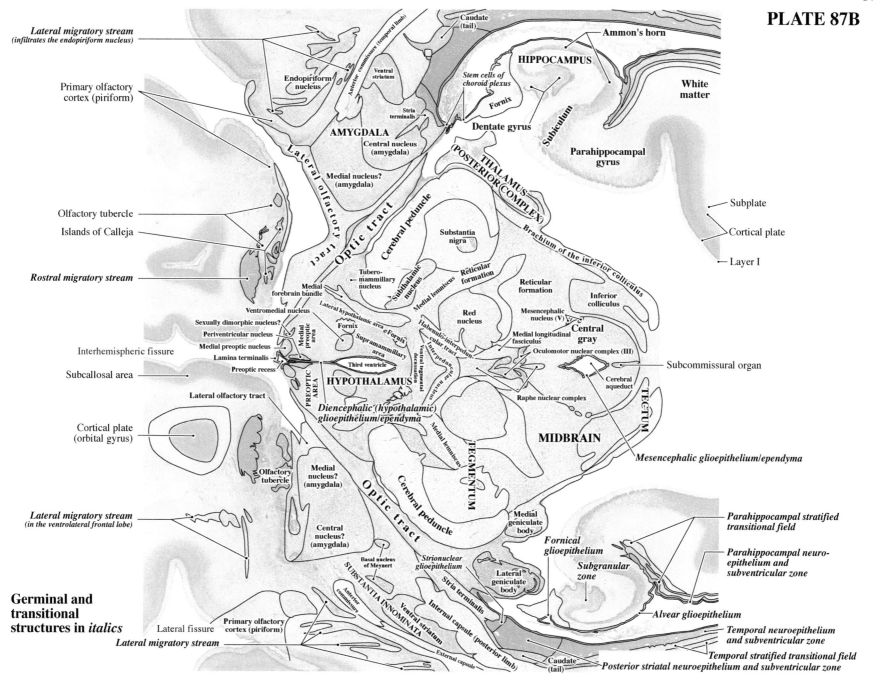

Lateral migratory stream
(infiltrates the endopiriform nucleus)

Caudate (tail)

Ammon's horn

HIPPOCAMPUS

Stem cells of choroid plexus

White matter

Ventral striatum

Anterior commissure (temporal limb)

Primary olfactory cortex (piriform)

Endopiriform nucleus

Fornix

Dentate gyrus

Subiculum

Stria terminalis

Parahippocampal gyrus

AMYGDALA

Central nucleus (amygdala)

THALAMUS POSTERIOR COMPLEX

Subplate

Medial nucleus? (amygdala)

Cortical plate

Olfactory tubercle

Substantia nigra

Brachium of the inferior colliculus

Layer I

Islands of Calleja

Subthalamic nucleus

Reticular formation

Reticular formation

Tubero-mammillary nucleus

Medial lemniscus

Inferior colliculus

Rostral migratory stream

Medial forebrain bundle

Lateral hypothalamic area

Mesencephalic nucleus (V)

Ventromedial nucleus

Fornix

Red nucleus

Central gray

Sexually dimorphic nucleus?

Fornix

Habenulo-interpeduncular tract

Medial longitudinal fasciculus

Periventricular nucleus

Supramammillary area

Oculomotor nuclear complex (III)

Medial preoptic nucleus

Interpeduncular nucleus

Interhemispheric fissure

Lamina terminalis

Medial preoptic area

Ventral tegmental decussation

Subcommissural organ

Subcallosal area

Preoptic recess

Third ventricle

Cerebral aqueduct

HYPOTHALAMUS

Raphe nuclear complex

TECTUM

Lateral olfactory tract

Diencephalic (hypothalamic) glioepithelium/ependyma

MIDBRAIN

Cortical plate (orbital gyrus)

Medial lemniscus

TEGMENTUM

Mesencephalic glioepithelium/ependyma

Olfactory tubercle

Medial nucleus? (amygdala)

Lateral migratory stream
(in the ventrolateral frontal lobe)

Cerebral peduncle

Medial geniculate body

Parahippocampal stratified transitional field

Central nucleus? (amygdala)

Fornical glioepithelium

Parahippocampal neuroepithelium and subventricular zone

Basal nucleus of Meynert

Strionuclear glioepithelium

Subgranular zone

Germinal and transitional structures in *italics*

Primary olfactory cortex (piriform)

SUBSTANTIA INNOMINATA

Stria terminalis

Lateral geniculate body

Alvear glioepithelium

Lateral fissure

Anterior commissure

Ventral striatum

Internal capsule (posterior limb)

Temporal neuroepithelium and subventricular zone

Lateral migratory stream

External capsule

Caudate (tail)

Temporal stratified transitional field

Posterior striatal neuroepithelium and subventricular zone

Optic tract

Cerebral peduncle

Lateral olfactory tract

**PLATE 88A
CR 260 mm
GW 30, Y187-65
Horizontal
Section 841**

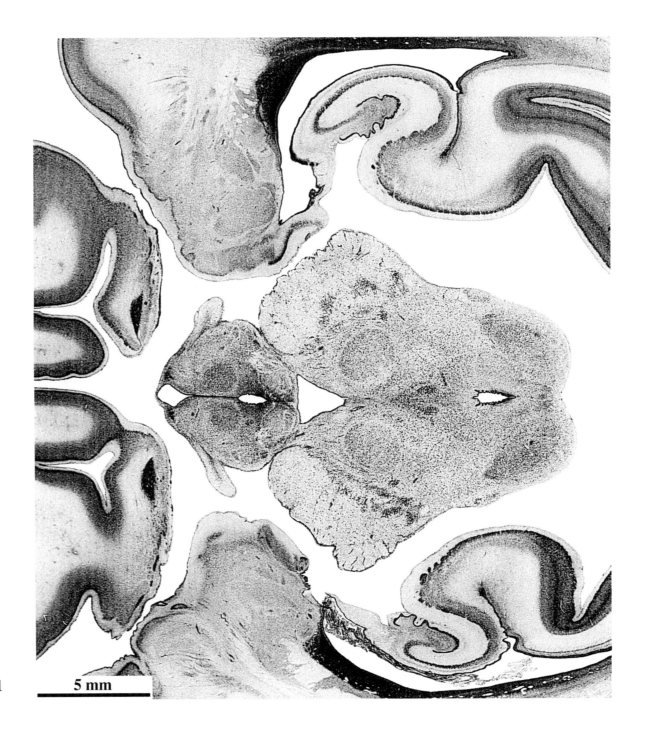

5 mm

**See the entire Section 841
in Plates 66A and B.**

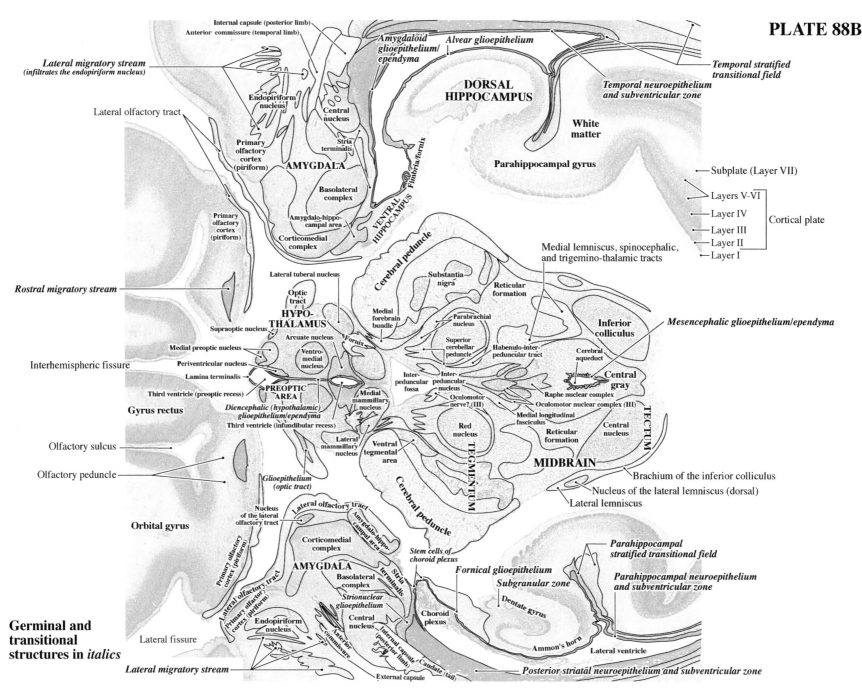

Internal capsule (posterior limb)

Anterior commissure (temporal limb)

Amygdaloid glioepithelium/ ependyma

Alvear glioepithelium

Temporal stratified transitional field

DORSAL HIPPOCAMPUS

Temporal neuroepithelium and subventricular zone

Lateral migratory stream (*infiltrates the endopiriform nucleus*)

White matter

Endopiriform nucleus

Central nucleus

Lateral olfactory tract

Parahippocampal gyrus

Primary olfactory cortex (piriform)

Stria terminalis

AMYGDALA

Subplate (Layer VII)

Basolateral complex

Fimbria/fornix

Layers V-VI

Layer IV

Cortical plate

Amygdalo-hippo-campal area

VENTRAL HIPPOCAMPUS

Layer III

Primary olfactory cortex (piriform)

Corticomedial complex

Layer II

Layer I

Medial lemniscus, spinocephalic, and trigemino-thalamic tracts

Lateral tuberal nucleus

Rostral migratory stream

Cerebral peduncle

Substantia nigra

Reticular formation

Mesencephalic glioepithelium/ependyma

Optic tract

Medial forebrain bundle

Parabrachial nucleus

Inferior colliculus

HYPO-THALAMUS

Supraoptic nucleus

Superior cerebellar peduncle

Habenulo-inter-peduncular tract

Cerebral aqueduct

Interhemispheric fissure

Medial preoptic nucleus

Arcuate nucleus

Ventro-medial nucleus

Fornix

Periventricular nucleus

Inter-peduncular fossa

Inter-peduncular nucleus

Raphe nuclear complex

Central gray

Lamina terminalis

Third ventricle (preoptic recess)

PREOPTIC AREA

Oculomotor nerve?.(III)

Oculomotor nuclear complex (III)

Medial longitudinal fasciculus

Central nucleus

TECTUM

Gyrus rectus

Diencephalic (hypothalamic) glioepithelium/ependyma

Third ventricle (infundibular recess)

Medial mammillary nucleus

Red nucleus

Reticular formation

Olfactory sulcus

Lateral mammillary nucleus

Ventral tegmental area

TEGMENTUM

MIDBRAIN

Brachium of the inferior colliculus

Olfactory peduncle

Glioepithelium (optic tract)

Cerebral peduncle

Nucleus of the lateral lemniscus (dorsal)

Lateral lemniscus

Orbital gyrus

Nucleus of the lateral olfactory tract

Lateral olfactory tract

Amygdalo-hippo-campal area

Parahippocampal stratified transitional field

Corticomedial complex

Stem cells of choroid plexus

Fornical glioepithelium

Parahippocampal neuroepithelium and subventricular zone

Primary olfactory cortex (piriform)

AMYGDALA

Basolateral complex

Subgranular zone

Lateral olfactory tract

Stria terminalis

Strionuclear glioepithelium

Dentate gyrus

Germinal and transitional structures in *italics*

Primary olfactory cortex (piriform)

Lateral fissure

Endopiriform nucleus

Central nucleus

Internal capsule (posterior limb)

Choroid plexus

Anterior commissure

Ammon's horn

Lateral ventricle

Lateral migratory stream

Caudate (tail)

Posterior striatal neuroepithelium and subventricular zone

External capsule

PLATE 89A
CR 260 mm, GW 30, Y187-65, Horizontal, Section 861

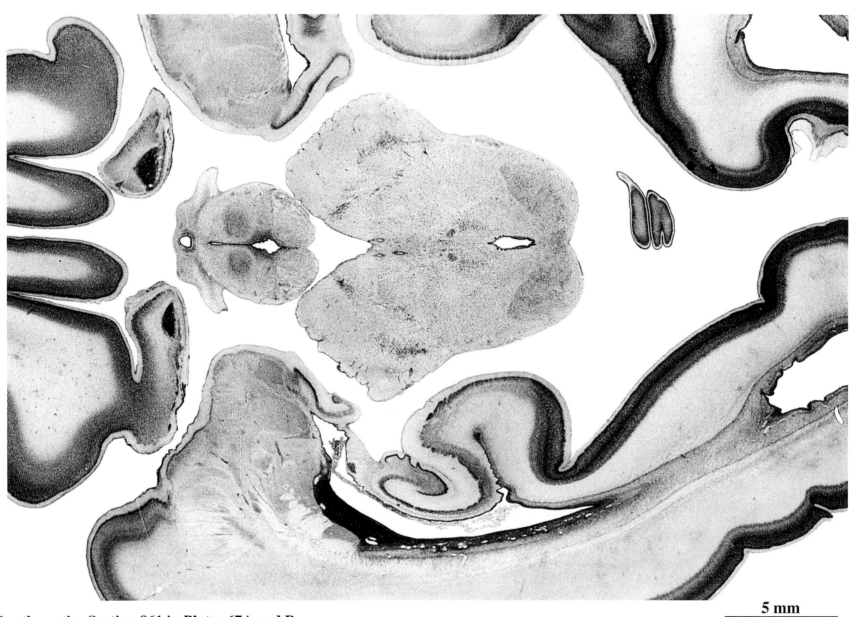

5 mm

See the entire Section 861 in Plates 67A and B.

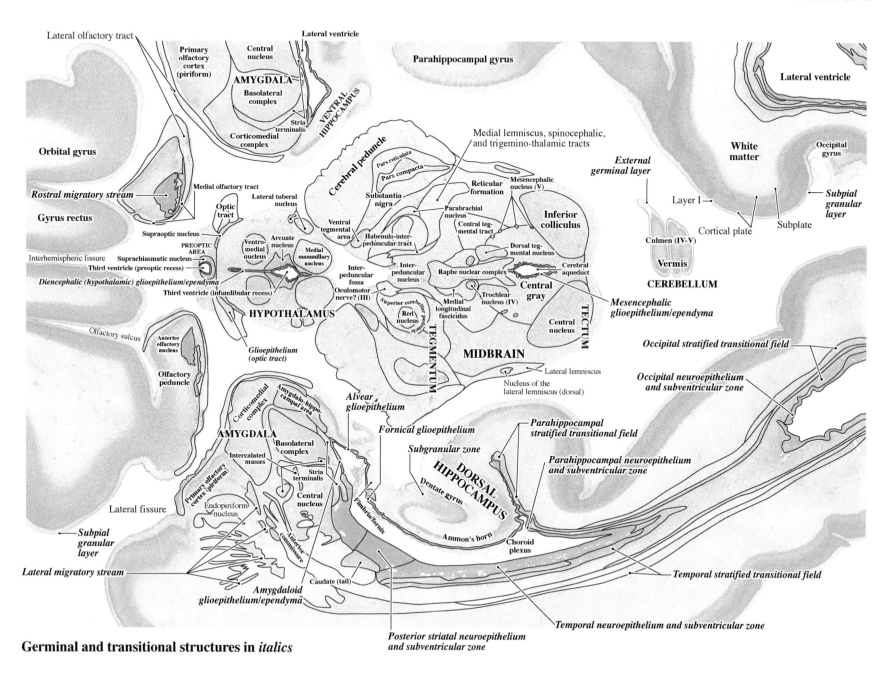

Germinal and transitional structures in *italics*

PLATE 90A
CR 260 mm, GW 30, Y187-65, Horizontal, Section 881

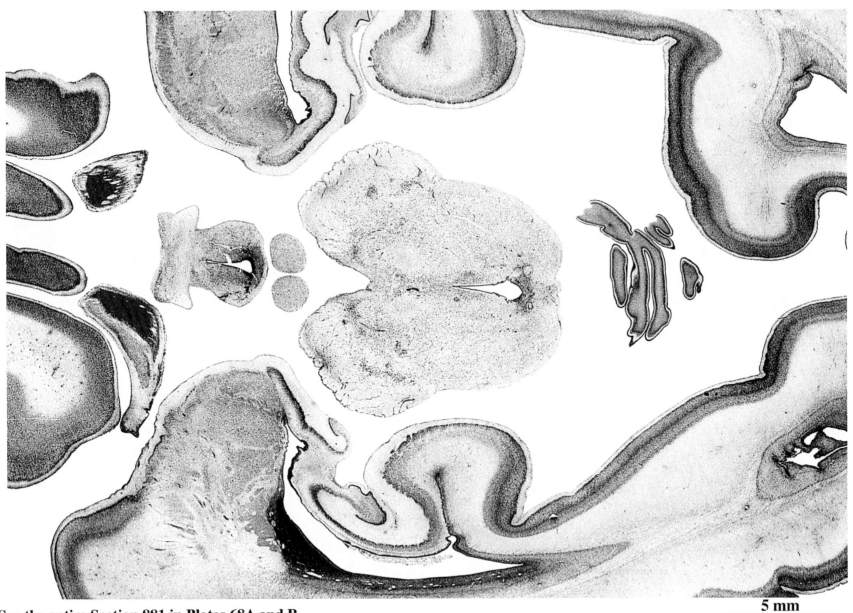

See the entire Section 881 in Plates 68A and B.

5 mm

Germinal and transitional structures in *italics*

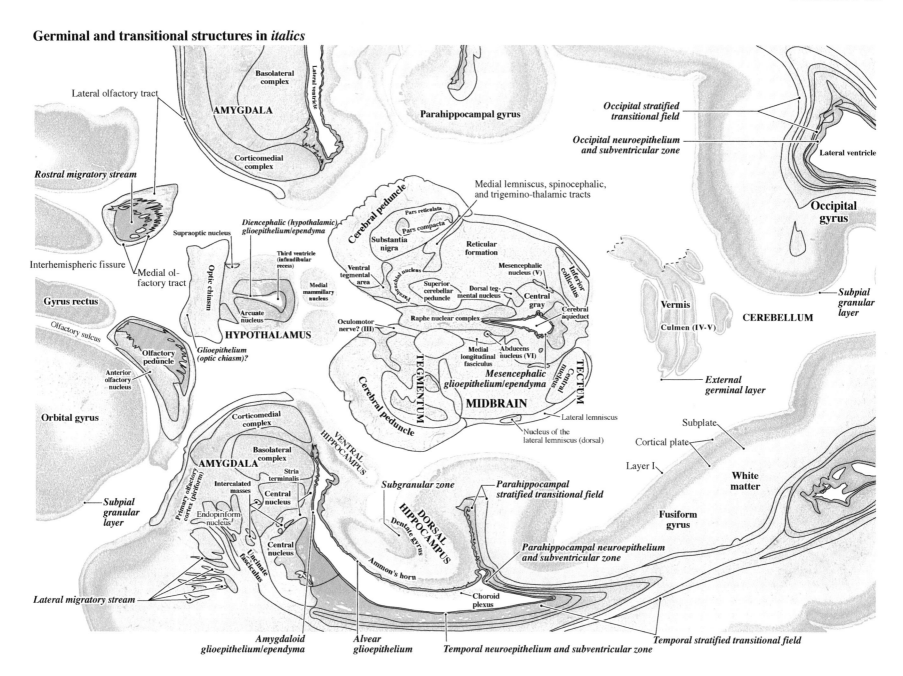

Basolateral complex

Lateral ventricle

Occipital stratified transitional field

AMYGDALA

Parahippocampal gyrus

Occipital neuroepithelium and subventricular zone

Lateral olfactory tract

Corticomedial complex

Lateral ventricle

Occipital gyrus

Rostral migratory stream

Cerebral peduncle

Medial lemniscus, spinocephalic, and trigemino-thalamic tracts

Pars reticulata

Pars compacta

Diencephalic (hypothalamic) glioepithelium/ependyma

Substantia nigra

Reticular formation

Inferior colliculus

Supraoptic nucleus

Third ventricle (infundibular recess)

Ventral tegmental area

Mesencephalic nucleus (V)

Interhemispheric fissure

Medial olfactory tract

Optic chiasm

Medial mammillary nucleus

Parabrachial nucleus

Superior cerebellar peduncle

Dorsal tegmental nucleus

Central gray

Subpial granular layer

Gyrus rectus

Arcuate nucleus

Raphe nuclear complex

Central nucleus

Cerebral aqueduct

Vermis

Olfactory sulcus

HYPOTHALAMUS

Oculomotor nerve? (III)

CEREBELLUM

Culmen (IV-V)

Olfactory peduncle

Glioepithelium (optic chiasm)?

Medial longitudinal fasciculus

Abducens nucleus (VI)

Anterior olfactory nucleus

TEGMENTUM

Mesencephalic glioepithelium/ependyma

Central nucleus

TECTUM

External germinal layer

Orbital gyrus

Cerebral peduncle

MIDBRAIN

Lateral lemniscus

Subplate

Cortical plate

Nucleus of the lateral lemniscus (dorsal)

Corticomedial complex

VENTRAL HIPPOCAMPUS

Layer I

White matter

Basolateral complex

Subgranular zone

Parahippocampal stratified transitional field

AMYGDALA

Stria terminalis

Intercalated masses

Central nucleus

Primary olfactory cortex (piriform)

Endopiriform nucleus

DORSAL HIPPOCAMPUS

Dentate gyrus

Parahippocampal neuroepithelium and subventricular zone

Fusiform gyrus

Subpial granular layer

Central nucleus

Uncinate fasciculus

Ammon's horn

Choroid plexus

Lateral migratory stream

Amygdaloid glioepithelium/ependyma

Alvear glioepithelium

Temporal neuroepithelium and subventricular zone

Temporal stratified transitional field

PLATE 91A
CR 260 mm, GW 30, Y187-65, Horizontal, Section 941

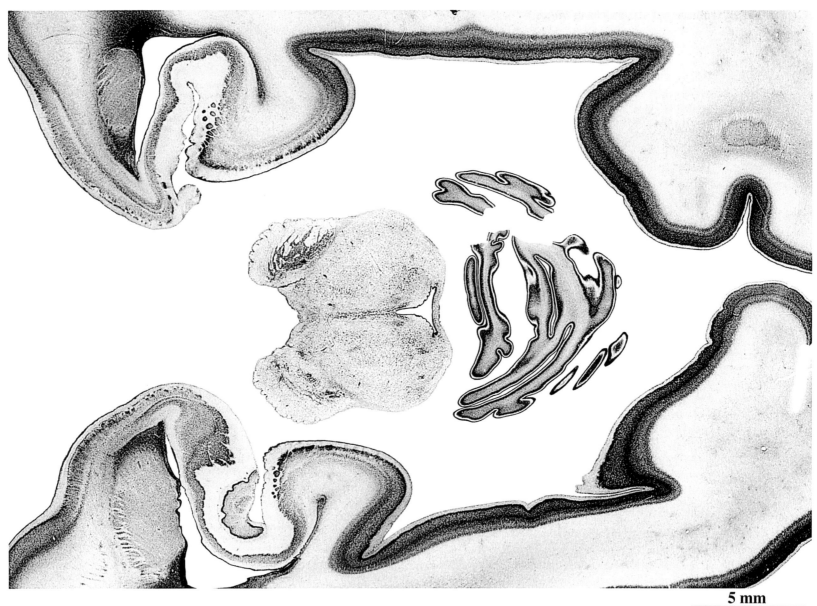

5 mm

See the entire Section 941 in Plates 69A and B.

Germinal and transitional structures in *italics*

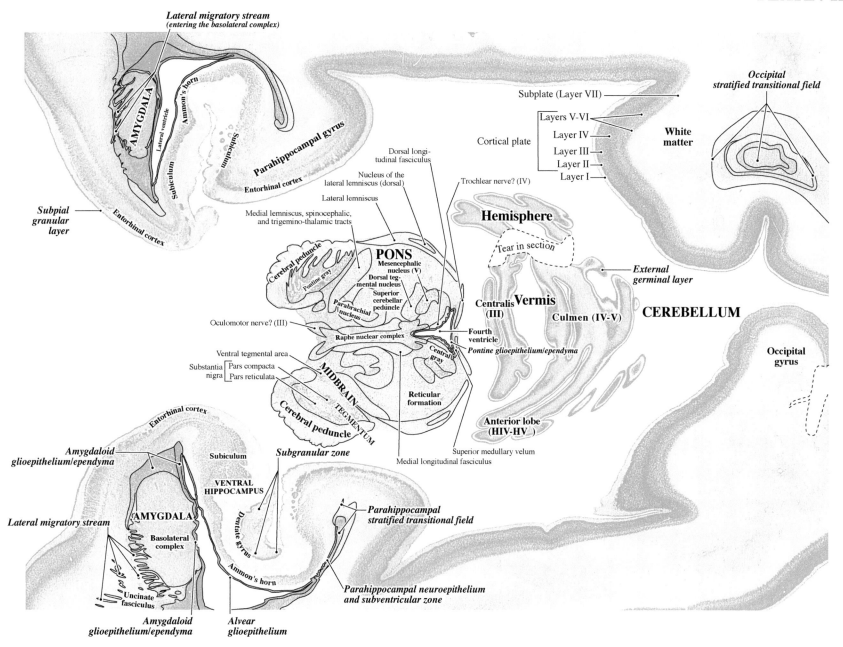

Lateral migratory stream
(entering the basolateral complex)

AMYGDALA

Lateral ventricle

Ammon's horn

Subiculum

Subiculum

Parahippocampal gyrus

Entorhinal cortex

Subpial granular layer

Entorhinal cortex

Subplate (Layer VII)

Layers V-VI

Cortical plate

Layer IV

Layer III

Layer II

Layer I

White matter

Occipital stratified transitional field

Dorsal longitudinal fasciculus

Nucleus of the lateral lemniscus (dorsal)

Trochlear nerve? (IV)

Lateral lemniscus

Medial lemniscus, spinocephalic, and trigemino-thalamic tracts

Hemisphere

Tear in section

External germinal layer

Cerebral peduncle

PONS

Mesencephalic nucleus (V)

Pontine gray

Dorsal tegmental nucleus

Superior cerebellar peduncle

Parabrachial nucleus

Centralis (III)

Vermis

Culmen (IV-V)

CEREBELLUM

Oculomotor nerve? (III)

Raphe nuclear complex

Fourth ventricle

Central gray

Ventral tegmental area

Pontine glioepithelium/ependyma

Substantia nigra — Pars compacta / Pars reticulata

MIDBRAIN

TEGMENTUM

Reticular formation

Occipital gyrus

Cerebral peduncle

Superior medullary velum

Medial longitudinal fasciculus

Anterior lobe (HIV-HV)

Entorhinal cortex

Amygdaloid glioepithelium/ependyma

Subiculum

VENTRAL HIPPOCAMPUS

Subgranular zone

Parahippocampal stratified transitional field

Lateral migratory stream

AMYGDALA

Basolateral complex

Dentate gyrus

Ammon's horn

Uncinate fasciculus

Amygdaloid glioepithelium/ependyma

Alvear glioepithelium

Parahippocampal neuroepithelium and subventricular zone

PLATE 92A
CR 260 mm, GW 30, Y187-65, Horizontal, Section 1001

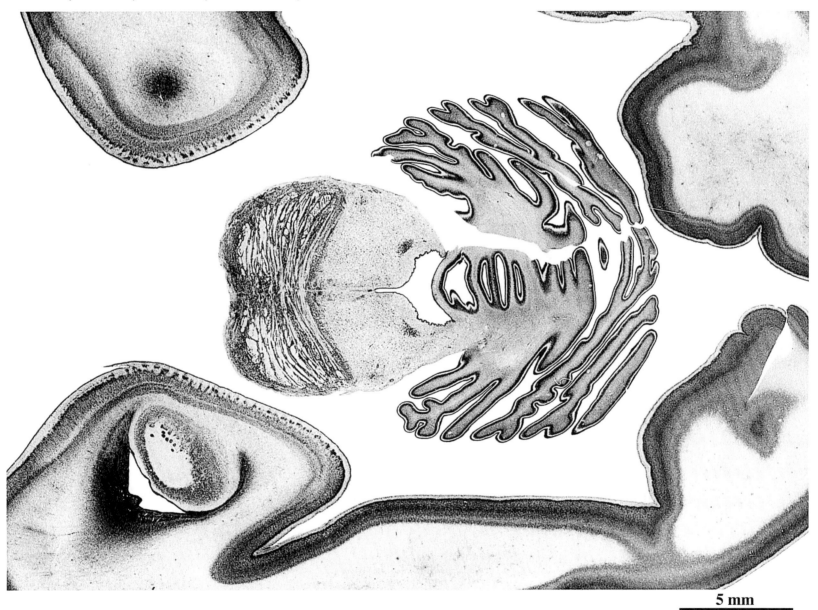

5 mm

See the entire Section 1001 in Plates 70A and B.

Germinal and transitional structures in *italics*

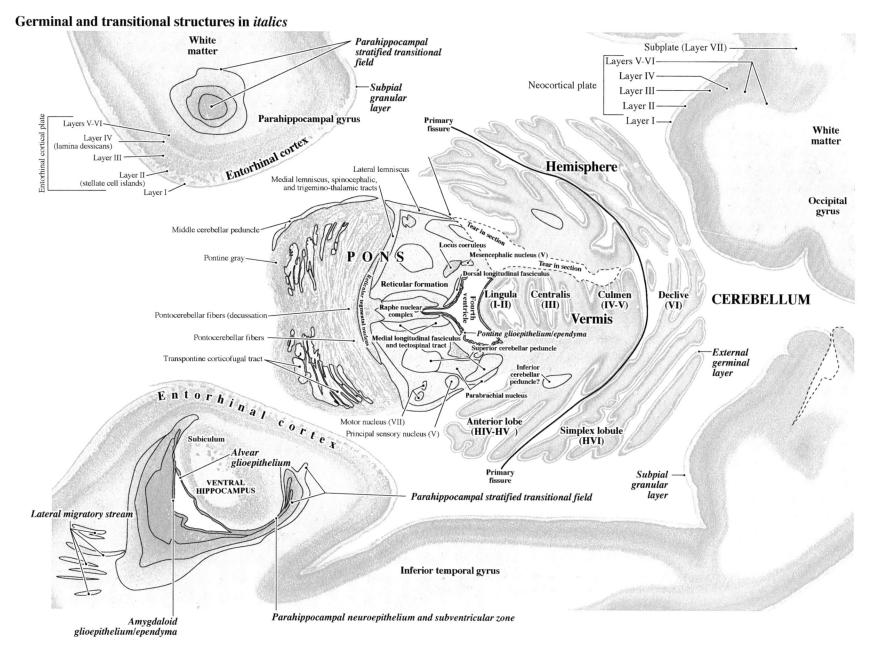

White matter

Parahippocampal stratified transitional field

Subpial granular layer

Parahippocampal gyrus

Entorhinal cortical plate

Layers V-VI
Layer IV (lamina dessicans)
Layer III
Layer II (stellate cell islands)
Layer I

Entorhinal cortex

Subplate (Layer VII)
Layers V-VI
Layer IV
Layer III
Layer II
Layer I

Neocortical plate

White matter

Occipital gyrus

Primary fissure

Hemisphere

Lateral lemniscus

Medial lemniscus, spinocephalic, and trigemino-thalamic tracts

Middle cerebellar peduncle

Pontine gray

Tear in section

Locus coeruleus

Mesencephalic nucleus (V)

Tear in section

P O N S

Dorsal longitudinal fasciculus

Reticular formation

Reticular tegmental nucleus

Pontocerebellar fibers (decussation

Raphe nuclear complex

Pontocerebellar fibers

Medial longitudinal fasciculus and tectospinal tract

Transpontine corticofugal tract

Fourth ventricle

Lingula (I-II)

Centralis (III)

Culmen (IV-V)

Declive (VI)

CEREBELLUM

Vermis

Pontine glioepithelium/ependyma

Superior cerebellar peduncle

Inferior cerebellar peduncle?

External germinal layer

Parabrachial nucleus

Motor nucleus (VII)

Principal sensory nucleus (V)

Anterior lobe (HIV-HV)

Simplex lobule (HVI)

Primary fissure

Subpial granular layer

E n t o r h i n a l   c o r t e x

Subiculum

Alvear glioepithelium

VENTRAL HIPPOCAMPUS

Lateral migratory stream

Parahippocampal stratified transitional field

Amygdaloid glioepithelium/ependyma

Parahippocampal neuroepithelium and subventricular zone

Inferior temporal gyrus

PLATE 93A
CR 260 mm
GW 30, Y187-65
Horizontal
Section 1041

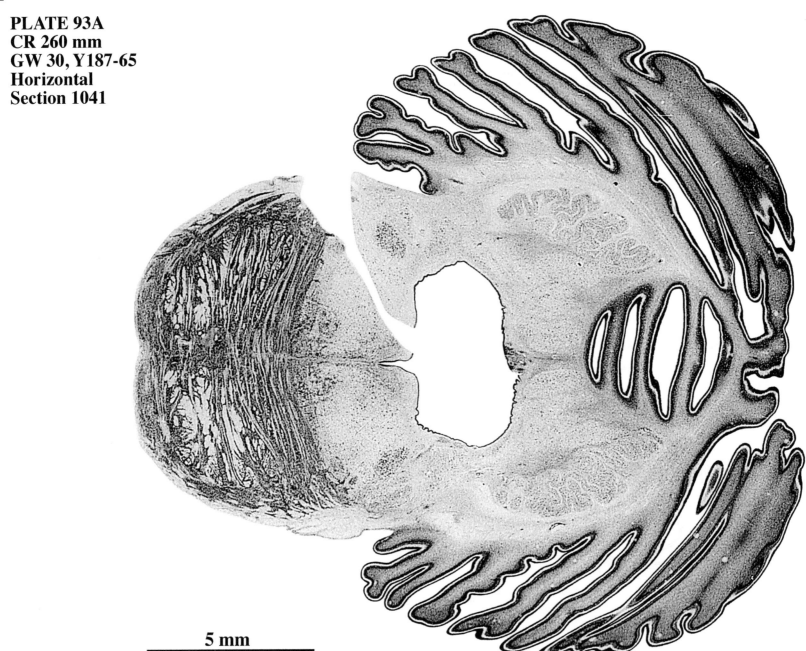

5 mm

See the entire Section 1041 in Plates 71A and B.

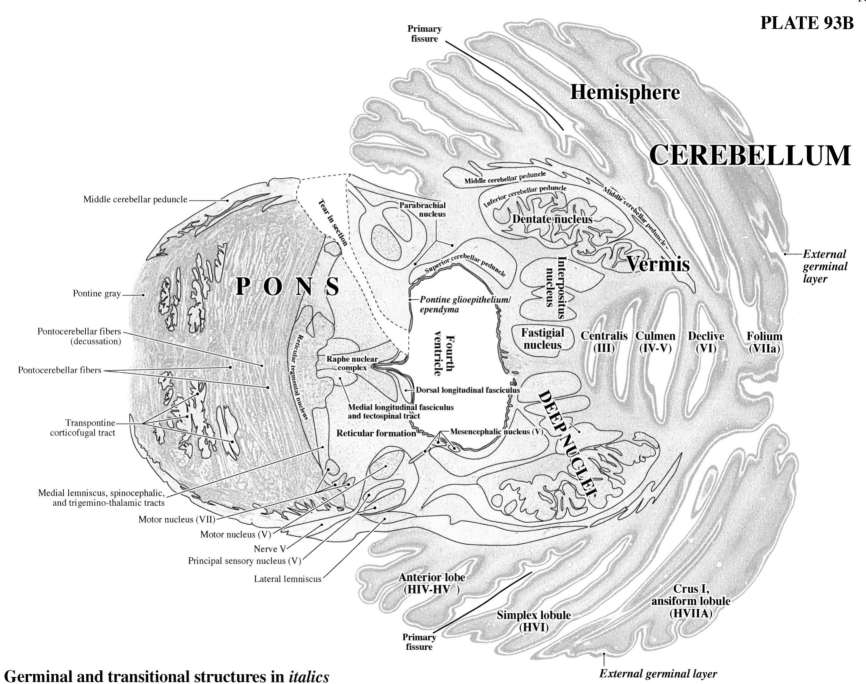

Primary
fissure

Hemisphere

CEREBELLUM

Middle cerebellar peduncle

Middle cerebellar peduncle

Inferior cerebellar peduncle

Parabrachial
nucleus

Tear in section

Dentate nucleus

Vermis

*External
germinal
layer*

Superior cerebellar peduncle

Interpositus
nucleus

Middle cerebellar peduncle

Middle cerebellar peduncle

P O N S

*Pontine glioepithelium/
ependyma*

Pontine gray

**Fourth
ventricle**

**Fastigial
nucleus**

**Centralis
(III)**

**Culmen
(IV-V)**

**Declive
(VI)**

**Folium
(VIIa)**

Pontocerebellar fibers
(decussation)

Reticular tegmental nucleus

Raphe nuclear
complex

Pontocerebellar fibers

Dorsal longitudinal fasciculus

DEEP NUCLEI

Medial longitudinal fasciculus
and tectospinal tract

Transpontine
corticofugal tract

Reticular formation

Mesencephalic nucleus (V)

Medial lemniscus, spinocephalic,
and trigemino-thalamic tracts

Motor nucleus (VII)

Motor nucleus (V)

Nerve V

Principal sensory nucleus (V)

Lateral lemniscus

**Anterior lobe
(HIV-HV)**

**Crus I,
ansiform lobule
(HVIIA)**

**Simplex lobule
(HVI)**

Primary
fissure

External germinal layer

Germinal and transitional structures in *italics*

PLATE 94A
CR 260 mm
GW 30, Y187-65
Horizontal
Section 1081

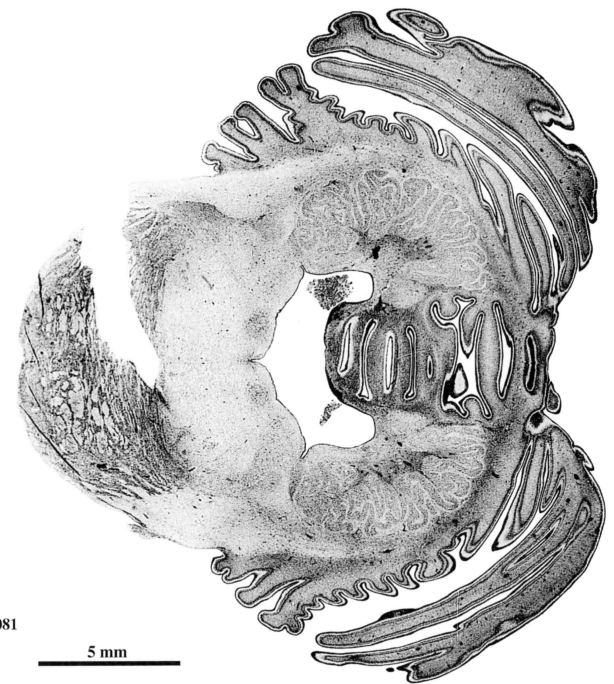

See the entire Section 1081
in Plates 72A and B.

5 mm

CEREBELLUM

Primary fissure

Simplex lobule (HVI)

Crus I, ansiform lobule (HVIIA)

Anterior lobe (HIV-HV)

Hemisphere

Middle cerebellar peduncle

Inferior cerebellar peduncle

Dorsal cochlear nucleus

Dentate nucleus

P O N S

Lateral lemniscus

Spinal nucleus and tract (V)

Tear in section

Superior olive complex

Vestibular nuclear complex

Superior cerebellar peduncle

Middle cerebellar peduncle

Motor nucleus (VII)

Nerve VII (genu)

Choroid plexus

Interpositus nucleus

External germinal layer

Pontine gray

Abducens nucleus (VI)

Germinal trigone

Vermis

Medial lemniscus and trapezoid body

Dorsal longitudinal fasciculus

Fourth ventricle

Nodulus (X)

Uvula (IX)

Pyramis (VIII)

Tuber (VIIa)

Folium (VIIa)

Pontocerebellar fibers (decussation)

Reticular tegmental nucleus

Raphe nuclear complex

Pontocerebellar fibers

Medial longitudinal fasciculus and tectospinal tract

Pontine glioepithelium/ependyma

Reticular formation

Transpontine corticofugal tract

Superior olive complex

Motor nucleus (VII)

Vestibular nuclear complex

Fourth ventricle (lateral recess)

Spinal nucleus and tract (V)

Superior cerebellar peduncle

Lateral lemniscus

Inferior cerebellar peduncle

DEEP NUCLEI

Dorsal cochlear nucleus

Middle cerebellar peduncle

Nerve V

Anterior lobe (HIV-HV)

Simplex lobule (HVI)

Crus I, ansiform lobule (HVIIA)

Primary fissure

External germinal layer

Germinal and transitional structures in *italics*

T - #0909 - 101024 - C51 - 210/279/9 - PB - 9781032228761 - Gloss Lamination